The Handbook
of Environmental Chemistry

Volume 2   Part C

Edited by O. Hutzinger

Reactions and Processes

With Contributions by
A.-S. Allard, E. F. King, A.W. Klein,
D. Mackay, A. H. Neilson,
H. A. Painter, S. Paterson, M. Remberger

With 49 Figures

Springer-Verlag Berlin Heidelberg GmbH 1985

Professor Dr. Otto Hutzinger

University of Bayreuth
Chair of Ecological Chemistry and Geochemistry
Postfach 3008, D-8580 Bayreuth
Federal Republic of Germany

ISBN 978-3-662-15261-4   ISBN 978-3-540-39048-0 (eBook)
DOI 10.1007/978-3-540-39048-0

Library of Congress Cataloging in Publication Data
Main entry under title: The Handbook of environmental chemistry.
Includes bibliographies and indexes.
Contents: v. 1, pts. A–C. The natural environment and the biogeochemical cycles/with contributions by P. Craig...
[et al.] – v. 2, pts. A–C. Reactions and processes/with contributions by W. A. Bruggeman... [et al.] – v. 3, pts. A–B.
Anthropogenic compounds/with contributions by R. Anliker... [et al.]
1. Environmental chemistry–Collected works. I. Hutzinger, O.
QD31.H335   574.5′222   80-16607

Preface

Environmental Chemistry is a relatively young science. Interest in this subject, however, is growing very rapidly and, although no agreement has been reached as yet about the exact content and limits of this interdisciplinary discipline, there appears to be increasing interest in seeing environmental topics which are based on chemistry embodied in this subject. One of the first objectives of Environmental Chemistry must be the study of the environment and of natural chemical processes which occur in the environment. A major purpose of this series on Environmental Chemistry, therefore, is to present a reasonably uniform view of various aspects of the chemistry of the environment and chemical reactions occurring in the environment.

The industrial activities of man have given a new dimension to Environmental Chemistry. We have now synthesized and described over five million chemical compounds and chemical industry produces about hundred and fifty million tons of synthetic chemicals annually. We ship billions of tons of oil per year and through mining operations and other geophysical modifications, large quantities of inorganic and organic materials are released from their natural deposits. Cities and metropolitan areas of up to 15 million inhabitants produce large quantities of waste in relatively small and confined areas. Much of the chemical products and waste products of modern society are released into the environment either during production, storage, transport, use or ultimate disposal. These released materials participate in natural cycles and reactions and frequently lead to interference and disturbance of natural systems.

Environmental Chemistry is concerned with *reactions in the environment*. It is about distribution and equilibria between environmental compartments. It is about reactions, pathways, thermodynamics and kinetics. An important purpose of this Handbook is to aid understanding of the basic distribution and chemical reaction processes which occur in the environment.

Laws regulating toxic substances in various countries are designed to assess and control risk of chemicals to man and his environment. Science can contribute in two areas to this assessment; firstly in the area of toxicology and secondly in the area of chemical exposure. The available concentration ("environmental exposure concentration") depends on the fate of chemical compounds in the environment and thus their distribution and reaction behaviour in the environment. One very important contribution of Environmental Chemistry to the above mentioned toxic substances laws is to develop laboratory test methods, or mathematical correlations and models that predict the environmental fate of new chemical compounds. The third purpose of this Handbook is to help in the basic understanding and development of such test methods and models.

The last explicit purpose of the Handbook is to present, in concise form, the most important properties relating to environmental chemistry and hazard assessment for the most important series of chemical compounds.

At the moment three volumes of the Handbook are planned. Volume 1 deals with the natural environment and the biogeochemical cycles therein, including some background information such as energetics and ecology. Volume 2 is concerned with reactions and processes in the environment and deals with physical factors such as transport and adsorption, and chemical, photochemical and biochemical reactions in the environment, as well as some aspects of pharmacokinetics and metabolism within organisms. Volume 3 deals with anthropogenic compounds, their chemical backgrounds, production methods and information about their use, their environmental behaviour, analytical methodology and some important aspects of their toxic effects. The material for volume 1, 2 and 3 was each more than could easily be fitted into a single volume, and for this reason, as well as for the purpose of rapid publication of available manuscripts, all three volumes were divided in the parts A and B. Publisher and editor hope to keep materials of the volumes one to three up to date and to extend coverage in the subject areas by publishing further parts in the future. Readers are encouraged to offer suggestions and advice as to future editions of "The Handbook of Environmental Chemistry".

Most chapters in the Handbook are written to a fairly advanced level and should be of interest to the graduate student and practising scientist. I also hope that the subject matter treated will be of interest to people outside chemistry and to scientists in industry as well as government and regulatory bodies. It would be very satisfying for me to see the books used as a basis for developing graduate courses on Environmental Chemistry.

Due to the breadth of the subject matter, it was not easy to edit this Handbook. Specialists had to be found in quite different areas of science who were willing to contribute a chapter within the prescribed schedule. It is with great satisfaction that I thank all 52 authors from 8 countries for their understanding and for devoting their time to this effort. Special thanks are due to Dr. F. Boschke of Springer for his advice and discussions throughout all stages of preparation of the Handbook. Mrs. A. Heinrich of Springer has significantly contributed to the technical development of the book through her conscientious and efficient work. Finally I like to thank my family, students and colleagues for being so patient with me during several critical phases of preparation for the Handbook, and to some colleagues and the secretaries for technical help.

I consider it a privilege to see my chosen subject grow. My interest in Environmental Chemistry dates back to my early college days in Vienna. I received significant impulses during my postdoctoral period at the University of California and my interest slowly developed during my time with the

National Research Council of Canada, before I could devote my full time to Environmental Chemistry, here in Amsterdam. I hope this Handbook may help deepen the interest of other scientists in this subject.

O. Hutzinger

Preface to Parts C of the Handbook

Parts C of the three series
– The Natural Environment and the Biogeochemical Cycles (Vol. 1)
– Reactions and Processes (Vol. 2)
– Anthropogenic Compounds (Vol. 3)
are now available. During their preparation it became obvious that further parts will have to follow to present the respective subject matters in reasonably complete form.

The publisher and editor have further agreed to expand the Handbook by three new series: Air Pollution, Water Pollution and Environmental Trace Analysis.

Again, I thank all authors as well as collaborators at the Springer Publishing House for their cooperation and help. Thanks are also due to many environmental chemists and reviewers in particular for their critical comments and their positive reception of the Handbook.

Bayreuth, December 1983   Otto Hutzinger

Contents

Biodegradation of Water-Soluble Compounds

H. A. Painter, E. F. King

The Fugacity Concept in Environmental Modelling
S. Paterson, D. Mackay

List of Contributors

Dr. Ann-Sofie Allard
Swedish Environmental
Research Institute
Box 21060
S-100 31 Stockholm
Sweden

Dr. E. F. King
Water Research Centre
Stevenage SG1 1TH
England

Dr. Albrecht W. Klein
Umweltbundesamt
Bismarckplatz 1
D-1000 Berlin 33

Prof. Dr. Donald Mackay
Dept. of Chemical Engineering
and Applied Chemistry
University of Toronto
Toronto, Canada M5S 1A4

Dr. Alasdair H. Neilson
Swedish Environmental
Research Institute
Box 21060
S-100 31 Stockholm
Sweden

Dr. H. A. Painter
Water Research Centre
Stevenage SG1 1TH
England

Dr. Sally Paterson
Dept. of Chemical Engineering
and Applied Chemistry
University of Toronto
Toronto, Canada M5S 1A4

Dr. Mikael Remberger
Swedish Environmental
Research Institute
Box 21060
S-100 31 Stockholm
Sweden

OECD Fate and Mobility Test Methods

A. W. Klein

Umweltbundesamt, Bismarckplatz 1,
D-1000 Berlin 33, Federal Republic of Germany

Introduction

The Organisation for Economic Cooperation and Development (OECD) started activities relating to chemicals control with special emphasis on new chemicals in 1974, when the Council urged governments of Member countries to make all efforts to ensure that the potential effects of chemicals and chemical products on man and his environment be assessed prior to marketing [1]. Following a broad study undertaken by the OECD Chemicals Group and the result of negotiations within the group, an assessment scheme was outlined to be used mainly in predicting the hazards of a new chemical [2]. The main features of this system can be summarised as follows:

(i)   when chemicals are subjected to systematic assessment, they should be considered in terms of hazard to *both* human health and the environment;

(ii) in the majority of cases, it is possible to determine no more than the likelihood of effects of chemicals on man or in the environment, and this can only be done through the application of expert judgement based on methods that are technically practicable as well as economically acceptable;

(iii) responsibility for generating and assessing the data necessary to determine the potential effects of chemical substances must be part of the overall function and responsibility of industry;

(iv) for the purpose of assessing the potential effects of a chemical substance and the likelihood that man and/or the environment may be exposed to such a substance, a *phased* approach should be applied.

In fact, as indicated under the last item, the hazard of a chemical is a function of two broad considerations:

– *the potential of the chemical to harm biological or other systems* and
– *the potential for its exposure without which, of course, the harm cannot occur* [3].

The first element is generally referred to as *effects* and covers both human and environmental effects. The second element has to consider the likelihood that a population or an object will be exposed to a chemical agent taking into account its *use patterns, releases,* and *environmental fate.* Generally used to mean a description of a chemical's transport and transformation behaviour in a selected environmental setting, environmental fate is principally a function of the chemical's *mobility,* the possibility of its being eliminated in the environment through *degradation/metabolization* processes and its potential for *accumulation* in specific sub-sectors of the environment. It therefore relies on

a) chemical inherent data, such as physical/chemical properties, degradation potential and accumulation potential, and
b) properties of the receiving environment, i.e. climatic, meteorological and geographical conditions [4].

Ranging from simple representations of the world to highly sophisticated, physics-based models, there are currently a number of approaches available for the prediction of a chemical's fate [5, 6]. All these approaches can, however, only be applied to a limited extent for new chemicals, because they are dependent on the data available, i.e. those data which are required under the respective legislation in each country.

The OECD Chemicals Testing Programme

In attempting to regulate test and evaluation requirements for chemicals with respect to their potential effects upon man and the environment, it must be realised that some 1,000 to 2,000 new chemical compounds enter the market per year and find world-wide usage. In addition, consideration must be given to those existing substances amongst the 50,000–60,000 already on the market which are suspected of being harmful. The nature of these substances and their designated use patterns vary greatly and they are found in many different combinations.

During the 1970s, therefore, as a number of key chemical-producing countries initiated their chemical control legislations, concern has been expressed that non-tariff barriers could be caused due to different and inconsistent requirements of test data or to a lack of mutual acceptance of these data. In 1977, in response to this concern, the OECD launched the "Chemicals Testing Programme" to make testing of chemicals more systematic, relevant and cost-effective with the aim of increasing exchange and acceptance of information between countries.

In order to prepare state-of-the-art reports on test methods that can be used to provide data on chemical substances for predictive purposes, five Expert Groups under the leadership of individual Member countries were established covering the following areas:

(i)   *physical-chemical properties* (Lead country-Germany)
(ii) *effects on biotic systems other than man* (Lead country – the Netherlands)
(iii) *degradation/accumulation* (Lead countries – Germany/Japan)
(iv) *long-term health effects* (Lead country – the United States)
(v)   *short-term health effects* (Lead country – the United Kingdom)

A sixth Expert Group, the Step Systems Group, under the leadership of Sweden drew upon the results of the other Expert Groups and worked on the tiered approaches recommended for testing and assessment of chemical hazard to man and his environment [2].

The five Expert Groups working on test methods took the following courses of action [7]:

- Selection of the most important properties of chemical substances within their respective areas of testing in relation to health and environmental hazard assessment.
- Assembling appropriate information from the OECD Member countries, national and international organisations and institutions, in particular information on:
  - officially codified standard methods,
  - consensus, i.e. well-known, methods (techniques) which are used routinely in different laboratories, but do not have official status as standard methods, and
  - methods from scientific literature.
- Analysis of the available material with respect to its
  - applicability to all kinds of chemical substances,
  - practicability (ease of performance),
  - cost (labour and equipment).
- Drafting of Test Guidelines containing descriptions of the methods to be used and guidance for the evaluation of results.
- Improvement of the drafted Test Guidelines
  - by circulating them among the scientific public of the different OECD Member countries for comment, and
  - where necessary and possible in the time-frame, by conducting international laboratory comparison studies.

After a two-year effort, final reports containing draft Test Guidelines as well as an analysis of approaches to testing within the respective areas could be presented. An extensive consultative process among the different Member countries followed and resulted in the adoption of official *OECD Guidelines for Testing of Chemicals* [8] under the OECD Council Decision on Mutual Acceptance of Data [9].

The decision affirms that data generated in one country according to the OECD Test Guidelines – in accordance with the *OECD Principles of Good Laboratory Practice* [9] – should be accepted in the Member countries for purposes of assessment and other uses relating to the protection of man and his environment. As a result, scarce resources are saved in both industrial and governmental administrations.

Test Parameters Relevant to Environmental Fate Assessment

To understand the behaviour of a chemical in the environment as completely as possible its fate analysis has to consider each of the potential environmental transport and transformation pathways. Due to the large number and variety of new chemicals entering the market world-wide, however, an economic and pragmatic approach is required which not only provides adequate protection of the environment but also enables costs to be minimised.

Ideally, a minimum fixed testing scheme and certain evaluation principles should be used which allow preliminary *screening* of chemicals behaviour and

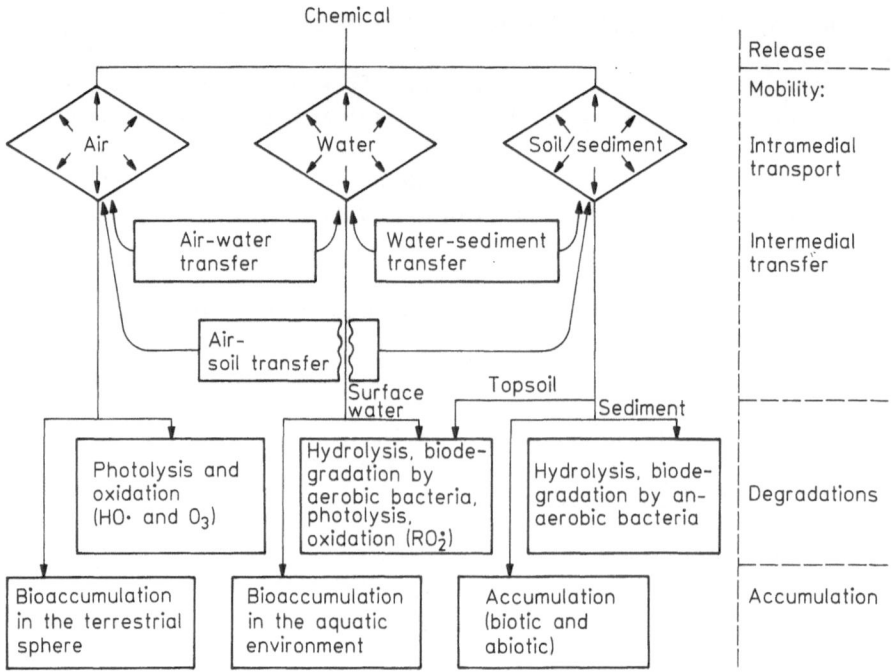

Fig. 1. Major pathways to be considered in environmental fate prediction

leave sufficient flexibility to accommodate more detailed studies in individual cases [10]. Since mobility and fate are governed as much by environmental characteristics as by the intrinsic properties of the chemical, this system should only take into account the major environmental pathways in order to maximise its general applicability to variations of the ecosystems under consideration and, thus, to provide estimates of general behaviour patterns.

The major pathways to be considered are sketched out schematically in Fig. 1 and can be classified into three categories:

● intramedial transport and intermedial transfer (mobility)
● degradation
● accumulation.

Properties of a chemical substance selected by the Expert Groups on physical-chemical properties and degradation/accumulation as most important in relation to these topics and for which official Test Guidelines were designed are listed in Table 1.

Mobility Test Methods

The mobility of a substance in the environment is determined by its tendency to partition between the different environmental media and its dispersion behaviour within each medium. Table 2 shows how chemical specific properties selected by the physical chemistry Expert Group can function for mobility assessment and

Table 1. List of OECD guidelines for determining physical-chemical properties and degradation/accumulation behaviour of chemicals

No	Subject

1. Physical-chemical properties

101	UV-VIS absorption spectra
102	Melting point/Melting range
103	Boiling point/Boiling range
104	Vapour pressure curve
105	Water solubility
106	Adsorption/Desorption
107	Partition coefficient (n-octanol/water)
108	Complex formation ability in water
109	Density of liquids and solids
110	Particle size distribution/Fibre length and diameter distributions
111	Hydrolysis as a function of pH
112	Dissociation constants in water
113	Screening test for thermal stability and stability in air
114	Viscosity of liquids
115	Surface tension of aqueous solutions
116	Fat solubility of solid and liquid substances

2. Degradation and accumulation

Ready Biodegradability

301 A	Modified AFNOR test
301 B	Modified Sturm test
301 C	Modified MITI test (I)
301 D	Closed bottle test
301 E	Modified OECD screening test

Inherent biodegradability

302 A	Modified SCAS test
302 B	Modified Zahn-Wellens test
302 C	Modified MITI test (II)

Simulation test – Aerobic sewage treatment

| 303 A | Coupled units test |

Biodegradability in soil

| 304 A | Inherent biodegradability test in soil |

Bioaccumulation

305 A	Sequential static fish test
305 B	Semi-Static fish test
305 C	Degree of bioconcentration in fish
305 D	Static fish test
305 E	Flow-Through fish test

what factors may affect the measured values or are operative under actual environmental conditions. Additionally various test methods are indicated which were identified to be included in the base set level of a phased (step sequence) testing and assessment approach for new chemical substances.

Table 2. Chemical specific properties relating to environmental mobility

Property	Function for mobility assessment	Affecting factors
Melting point/ Melting range[a]	Determines physical state during the life cycle of a chemical substance and relates therefore to its spreading potential; *base set property* for identification purposes	Impurities
Boiling point/ Boiling range[a]	Determines physical state during the life cycle of a chemical; relates to vapour pressure and indicates thus volatility; *base set property* for identification purposes	Impurities; ambient vapour pressure
Vapour pressure curve[a]	Relates to the rate of evaporation; input for prediction of volatility from water and soil and of air/water partitioning	Impurities
Water solubility[a]	Determines affinity for aqueous medium and affects largely the spatial and temporal movement of a chemical within and between air, soil and water compartments (hydrologic cycle)	Method of measurement, temperature, salinity of water, dissolved organic matter, pH
Adsorption/ Desorption[a]	Used to estimate soil (sediment)/water partitioning and leaching	Soil/sediment: organic carbon content, clay content, particle size distribution and surface area, cation exchange capacity, pH, temperature Water: pH, temperature, salinity, concentration of dissolved organic matter, suspended particulates. Test specific: nonequilibrium adsorption mechanisms or failure to reach equilibrium conditions, solids to solution ratio, loss of chemical due to volatilization, degradation, adsorption on glass walls, nonlinear isotherm.
Complex formation ability in water	Transformation process at the water/ soil-sediment interface leading to remobilization of metals present in soil and sediments	Temperature, pH, naturally occurring ligands such as humic and fulvic acids.
Density of liquids and solids[a]	Relates to relative distribution within and between the compartments water, soil, air; determining factor in the settling behaviour of water-insoluble liquids and solids	
Particle size distribution[a]	Determines tendency to become suspended in air and water and relates to settling out; influences the distribution and mass transport of insoluble and non-volatile particles in water, air and, in some cases, in the upper soil layers	Method of measurement (data from sedimentation methods in solvents often show lower particle sizes than corresponding data from dry (air) sedimentation).

Table 2 (continued)

Property	Function for mobility assessment	Affecting factors
Dissociation constants in water[a]	Affects transfer processes from the aquatic environment to the atmosphere and sediment or soil	Ionic strength of the aqueous medium, temperature
Viscosity of liquids	Affects the extent of spreading on land and in water and the penetration of fluids into soil	Temperature
Surface tension of aqueous solutions	Affects spreading on land and water, penetration into soil. Also important with respect to emulsification of liquids that are mixed with water and in their adsorption to solid surfaces	Temperature

[a] Base set property

Degradation Testing

Degradation has been selected as the term to describe any process in which the complexity of a chemical substance is reduced by the formation of simpler molecules. It includes abiotic degradation by chemical transformations such as photolysis (direct and indirect), free radical oxidation and hydrolysis as well as biotic degradation processes by aerobes and anaerobes. The OECD Expert Group on degradation/accumulation has selected topics for detailed review on biodegradation in water, abiotic degradation and degradation in soils and sediments.

Biodegradation in Water

As pointed out earlier, in order to assess a large number of new chemicals a system is required which allows their preliminary screening, using relatively simple tests, and the identification of those substances for which more detailed studies are needed. On the basis of a hierarchy of tests of increasing complexity and cost, the degradation/accumulation Expert Group organized the examination of biodegradability into a logical series of steps as follows:
(i)   ready biodegradability
(ii) inherent biodegradability
(iii) simulation tests.

Definitions

Ready biodegradability is the ability of a chemical to break down in stringent tests which provide limited opportunity for biodegradation and acclimatisation to occur. It may be assumed that a chemical giving a positive result in a test of this type will undergo rapid and ultimate biodegradation in the environment and, therefore, should be classified as "readily biodegradable".

Ultimate biodegradation is the breakdown of an organic chemical to carbon dioxide, water, the oxides or mineral salts of any other elements present, and products associated with the normal metabolic processes of micro-organisms.

Inherent biodegradability is the ability of a chemical to break down under more favourable test conditions than those provided in the ready biodegradation tests (acclimatisation, prolonged exposure, and a more favourable chemical/biomass ratio). A chemical giving a positive result in a test of this type may be classified as "inherently biodegradable", but, due to the favourable test conditions employed, it may not be assumed to biodegrade rapidly in the environment in every case.

Simulation tests are tests which provide evidence of the rate of biodegradation under some environmentally relevant conditions. Tests of this type may be subdivided according to the environment they are designed to simulate: a) biological treatment (aerobic); b) biological treatment (anaerobic); c) river; d) lake; e) estuary; f) sea; and g) soil.

Tests of Ready Biodegradability

Tests at this level must be designed so that positive results for ready biodegradability are unequivocal, because this evidence will mean that further examination of the chemical's biodegradability or of possible environmental effects of biodegradation products is normally required only under certain circumstances (special cases of use or exposure; high marketing levels; indications for high biological activity).

Five tests listed in Table 3 have been specified to determine the "ready biodegradability" of organic chemicals. The procedures are similar in that in all cases the test compound, which provides the sole source of organic carbon, is added to an aqueous solution of mineral salts and exposed to relatively low numbers of bacteria under aerobic conditions for a period of up to 28 days. To follow the course of biodegradation, a non-specific analytical method (e.g. loss of dissolved organic carbon (DOC), evolution of CO_2 or oxygen consumption) is used which has the advantage that the methods are immediately applicable to a wide variety of organic compounds without it being necessary to develop specific analytical procedures. Since these methods also respond to any biodegradation residues or intermediates, an indication of the extent of ultimate biodegradation is given.

Although differences in test conditions (e.g. inoculum, test concentration etc.) mean that chemicals may give different results in the various procedures, the *pass levels* as shown in Table 3 have been set so as to ensure that all chemicals meeting them can be classified as "readily biodegradable" regardless of which test method is used. Since the differences in test conditions arise in the main from the constraints imposed by the distinctive analytical method utilized, it is not possible to recommend one procedure which has general applicability to all organic chemicals. Particularly careful consideration must be given to the selection of the appropriate method for testing of chemicals which are likely to be toxic to bacteria, not soluble in water at the required concentrations or which volatilize. For ex-

Table 3. Pass levels for ready biodegradability tests

Test	Test duration (d)	Criterion	Pass level %	Suitable for poorly soluble substances	Suitable for volatile substances
Modified AFNOR test	28	DOC[a]	70		
Modified Sturm test	28	CO_2	70		
Modified MITI test (I)	28	BOD[b]	60	+	+
		LPC[c]	70		
Closed bottle test	28	BOD[b]	60	+	+
Modified OECD screening test	28	DOC[a]	70		

[a] Dissolved organic carbon, i.e. the amount of carbon present in organic molecules in aqueous solutions after removal of particulate matter including biomass
[b] Biochemical oxygen demand, i.e. the amount of oxygen consumed by microorganisms when metabolizing on substrate
[c] Loss of parent compound

ample, only the Closed Bottle Test and the modified MITI Test may be used for volatile substances whereas the modified OECD Screening Test and the AFNOR Test should *not* be used for chemicals which are poorly soluble or likely to adsorb strongly. On the other hand those methods based on DOC measurement do not have the advantage that both the mineralization *and* the conversion to biomass of the test compound are detected.

The main purpose in setting the test duration at 28 days was to allow sufficient time for adaption of the organisms to the chemical (lag phase). In order to avoid that compounds will pass the test, which degrade only slowly after a relatively short adaption period, the full biodegradation curve should be considered. In this way the duration of the lag phase, slope and plateau level can be identified and a check of the biodegradation rate can be made. Thresholds indicated in Table 3 must be reached within 10 days ("time window") beginning with the day on which 10 per cent biodegradation is observed.

Because of the stringency of the test procedures lack of ready biodegradation does not, however, necessarily mean that the test compound is not biodegradable at all under environmental conditions, but indicates that more work will be needed to establish biodegradability.

Tests of Inherent Biodegradability

Three tests, listed in Table 1, applying optimum conditions (relatively high concentrations of microorganisms, prolonged time periods) were designed to assess if a chemical has any potential for biodegradation. A negative result at this stage will therefore normally mean that further work on biodegradability is not necessary and that, in estimating the likely environmental fate of the chemical, non-biodegradability should be assumed.

On the other hand, a positive result indicates that the chemical will not persist indefinitely in the environment and, depending on the required accuracy of environmental fate prediction, it may be necessary to proceed to the next level.

Appropriate Simulation Test

At present experience with simulation testing is limited mainly to the simulation of sewage treatment. For this reason only one official OECD Test Guideline in this field has been designed concerning aerobic sewage treatment.

The principle of the test method is that two models of activated sludge plants are operated in parallel, the parallelism being increased and assured by a trans-inoculation procedure. The test material is added to the influent (synthetic sewage) of one unit while the other is fed only with the synthetic sewage. The dissolved organic carbon, DOC, (or chemical oxygen demand, COD) difference of these effluent values then is due to non- or only partially degraded test material.

The development and validation of methods which are models for ecosystems, such as rivers, shallow and deep lakes, estuaries, seas and soils is a subject of high priority. It has to be stressed, however, that simulation tests will, by their nature, in most cases be research projects. More insight in the kinetics of the degradation processes will be necessary to develop simpler test systems for quantitative information on rates of degradation directly relevant to a particular environmental situation.

Abiotic Degradation

The chemical processes of photolysis, free radical oxidation, and hydrolysis can be major routes for elimination of man-made chemicals from the environment and may occur in any medium, depending on the reactivity of the substance and the presence of the substrates capable of inducing chemical transformation.

Photochemically induced processes are of paramount importance for the removal of waste chemical substances from the atmosphere. Even when liquid phase hydrolysis is rapid, gas phase hydrolysis may be very slow. However, in 1979 when the Expert Group on degradation/accumulation presented its Final Report, test methods were few in number and none had been validated.

Meanwhile, a reasonable protocol for measuring atmospheric rate constants consisting of estimation methods to determine the dominant process(es) and laboratory measurements of the rate constants for dominant process(es) has been suggested [11] and is given in Fig. 2. The reactions considered include direct photochemical reaction and reactions with the OH radical and ozone. Physical processes are not taken into consideration, but these are expected to be significant for only a small number of chemicals [12].

The first tier of tests screens these three pathways and estimates the approximate significance of each. An upper limit for the rate of photolysis is obtained by assuming that reaction occurs with a quantum efficiency of unity and by equating the rate of reaction with the rate of absorption of light. The latter rate is determined by combining ultraviolet (UV)-visible absorption spectral data with the solar spectral intensities. A method of integrating the absorption and solar intensity data for various times of the year is provided. This method is relatively simple, yet has been shown to yield accurate direct photolysis constants [13].

Estimates of the rate constants for reaction with OH and O_3 are based on SAR developed by Hendry and Kenley [14]. From these rate constants and estimates

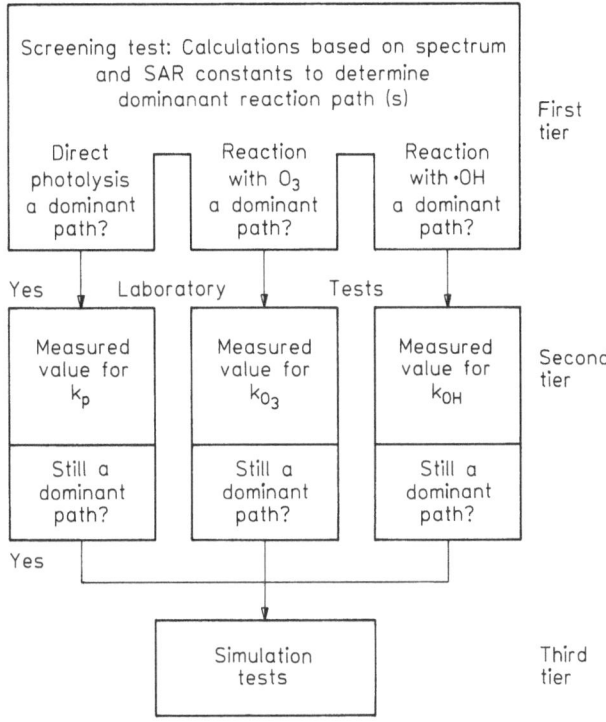

Fig. 2. Proposed protocol for measuring atmospheric rate constants

of the environmental concentrations of OH and O_3, the first order rate constant can be estimated for each process. These can then be directly compared to determine the relative significance of each process.

Once the relative significance of the three processes is determined, the rate constants for those processes which clearly dominate should be measured in detailed tests constituting the second tier of investigation.

Simulation testing in the third tier using large smog chambers or Tedlar (fluorocarbon) bags located outdoors requires more time and effort and is more prone to error than are laboratory measurements. They are, however, essential to gaining a widespread acceptance of the simpler, less "realistic" laboratory tests and estimation methods.

Degradation in Soil and Sediments

The study of the fate of chemicals in soils or sediments is of particular importance since contamination can be long lasting and difficult to reverse due to limited opportunity for mobilisation and dilution. Only in such test systems which include the soil and sediment in a way that leaves their structure functionally undisturbed, e.g. by using soil monoliths or field experiments, can the actual fate and behaviour of chemicals be investigated.

When performing tests for biodegradation, however, only the texture of the medium is normally maintained, whilst in almost all tests the naturally developed soil or sediment structure is destroyed, e.g. by sieving or washing. In order to make the results obtained intercomparable, the use of three standardized soil types according to the American soil classification scheme [15] has been recommended:

(1) Alfisol: pH between 5.5 and 6.5

             organic carbon content between 1 and 1.5 per cent

             clay content (i.e. particles <0.002 mm in diameter) between 10 and 20 per cent

             cation exchange capacity between 10 and 15 mval

(2) Spodosol: pH between 4.0 and 5.0

             organic carbon content between 1.5 and 3.5 per cent

             clay content ≤ 10 per cent

             cation exchange capacity < 10 mval

(3) Entisol: pH between 6.6 and 8.0

             organic carbon content between 1 and 4 per cent

             clay content between 11 and 25 per cent

             cation exchange capacity > 10 mval

All three are common in temperate zones but are not representative of arid or tropical zones. Since they provide sufficient flexibility and vary significantly in parameters such as cation exchange capacity, clay content, organic carbon content, exchangeable cations and pH, these soil types are also recommended for use in adsorption/desorption testing to consider the environmental characteristics affecting this property as indicated in Table 2.

As soil is a more powerful matrix for biodegradation testing than water, the need for a preliminary soil degradation test, similar to those for ready biodegradability in water, will be restricted to exceptional or localized cases, such as where chemicals are directly applied to soils. Thus, the procedure for soil/sediment testing will start at a level similar to the inherent biodegradability tier for aquatic conditions.

An appropriate test method on the "Inherent Biodegradability in Soil" has been included in the offical OECD Guidelines for Testing of Chemicals [8] (cf. Table 1). Since carbon dioxide evaluation in soil systems is not sensitive enough to be used as a criterion for accomplished degradation, the test is based on ^{14}C-labelled materials.

Simulation tests related to soil and sediments, again, will be research projects and should aim at analysis of the parent compounds and identification of its degradation products. For this purpose the use of soil microcosm studies with undisturbed soil-sediment conditions [16–18] is most appropriate. Specific chemical analysis may be utilized, but in most cases labelled chemicals will be better.

Accumulating Testing

Chemicals entering the environment generally disperse unevenly in and between the compartments water, soil, and air. This behaviour is largely determined by their physical and chemical properties and by the nature of the abiotic and biotic components of the environment. Key parameters for the rate of the corresponding transfer processes are, on the part of the chemical, water solubility, volatility, vapour pressure and sorptivity, and, on the part of the environment, phase structures, textures, and compartmented biological structures that provide an extreme variety of sites for dissolution and binding. Due to the dynamic nature, this partitioning behaviour of a given chemical results in the formation of concentration gradients which vary with time. Such a substance therefore "accumulates" (concentrates) for example in an abiotic phase (e.g. water→soil), in an organism (e.g. water→fish), or a tissue (e.g. blood→fatty tissue).

This accumulation of chemicals in the environment is a normal event. The extent of retention of a chemical, however, is essentially influenced by various factors, such as sorption and its susceptibility to metabolic transformations. Generally, the metabolites formed have physical and chemical properties different from those of the parent chemical, and they can again partition.

Accumulation in Soils and Sediments

Empirically it has been found that the partition coefficient (n-octanol/water) (Test Guideline 107, cf. Table 1) for nonionic organic chemicals can be correlated to the leaching characteristics [19] and the soil-sediment/water partitioning [20–22] as a measure of accumulation tendency in soils and sediments. If confirmatory tests are necessary, direct measurements of the adsorption coefficient between water and soil using Test Guideline 106 (see Table 1) can be made. For ionic materials the n-octanol/water partition coefficient is not a reliable indicator for accumulation in soil or sediment, and direct measurements of the sorption coefficient are required.

Bioaccumulation

The actual extent of bioaccumulation of a chemical at a given time is the combined result of the competing processes of uptake, distribution, transformation and excretion as sketched in Fig. 3 [23].

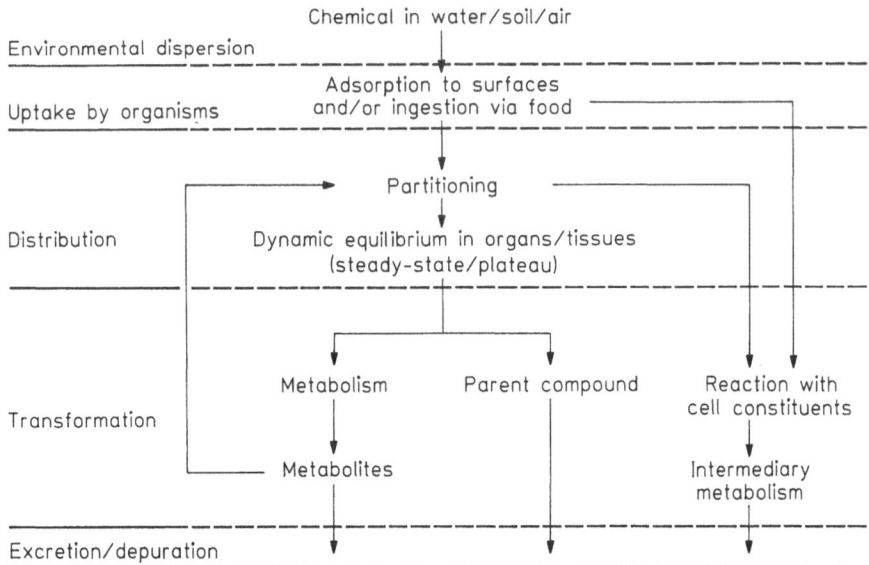

Fig. 3. Uptake and retention of chemicals

If the chemical under consideration is available for uptake by the organism over a period of time, the four interconnected phases result in a dynamic equilibrium (apparent plateau or steady state) which is characterized by a constant ratio of the concentration of the chemical in the test organism and in the ambient medium (e.g. water) or in food. This ratio is the mathematical quantification of bioaccumulation.

Two distinct and separable routes of uptake by organisms are commonly described:
1) *direct bioaccumulation* from the ambient medium, and
2) *indirect bioaccumulation* via the food chain *(biomagnification)*

Direct bioaccumulation unequivocally predominates over indirect bioaccumulation in the aquatic environment [24] whereas biomagnification is the dominant mechanism effective in terrestrial animals [25].

Bioaccumulation in the Aquatic Environment

There is a clear correlation between direct bioaccumulation in aquatic animals and the n-octanol water partition coefficient (P_{ow}) of non-ionised organic chemicals [21, 26]. Although exceptions are known, they relate to cases where P_{ow} leads to an overestimation of the corresponding bioconcentration factor (BCF), i.e. the ratio of the chemical's concentration in the whole organism to the concentration in the (test) environment under steady-state conditions. Therefore, the partition coefficient, P_{ow}, may be regarded as a good, conservative predictor of bioaccumulation of these chemicals in the aquatic environment.

For this reason, tests for bioaccumulation in fish are recommended only when $\log P_{ow} > 3$. When $\log P_{ow}$ is below 3, the estimated bioconcentration factor (wet weight basis), BCF, can be assumed to be below 100, and no further bioaccumulation test normally should be necessary. The same is true if a chemical is readily biodegradable, as defined under "Biodegradation in Water".

There might be cases, however, where the reaction of a chemical in a living organism may result in derivatives which are more lipophilic than the parent chemical (e.g. inorganic mercury *vs* methyl mercury). Such a reaction may also lead to binding of the chemical or parts of it to cell constituents. In the former case, bioaccumulation of the derivative will be higher than expected from the P_{ow} of the parent compound. In the latter case, the retention of the bound chemical will depend upon the stability of the bond in question and the metabolic half-life of the cell constituent the chemical is bound to.

Possible candidates for such a behaviour may be found among inorganic and organo-metallic chemicals. In these cases the bioaccumulation potential cannot be established unequivocally by the n-octanol/water partitioning test. Such compounds should undergo a bioaccumulation test on fish which at this screening stage would be one of the static tests listed in Table 1.

For ionic compounds that ionize under physiological conditions (pH 3–9) the use of the n-octanol water partition coefficient is also only predictive if the impact of complex and ion pair formation upon the bioaccumulation potential can be evaluated and evidence exists that the chemical will show no reactivity to cell constituents.

If this evidence cannot be provided, at the screening stage a static fish test should be performed for the parent compound or its products of rapid transformation unless the compound is known to be readily biodegradable.

Terrestrial Environment

Correlations between the n-octanol water partition coefficient, P_{ow}, and the bioconcentration factor for terrestrial organisms have not yet been established. However, studies on mammals and birds have confirmed that lipophilicity of a chemical of sufficient metabolic stability is the property which essentially determines the degree of bioaccumulation. Consequently, at the screening stage the n-octanol water partition coefficient is also a good, although rather qualitative predictor of the potential of a non-ionised organic chemical to bioaccumulate in lipid tissues of terrestrial animals. As their uptake of chemicals is mainly by food, species at the top of food chains can be expected to show the highest concentrations of bioaccumulative chemicals that enter such webs at any lower level.

In plants no significant bioaccumulation occurs, so that testing is not appropriate.

The OECD Minimum Premarketing Set of Data

Notwithstanding any pragmatic regulatory reasons for establishing sequential approaches to testing of new chemicals, it is well established scientific logic to proceed step by step in any complex laboratory testing programme. Before considering degradation testing, for example, water solubility and volatility should be considered in order to select the most appropriate test and, certainly, the performance of a fish accumulation study should in most cases be preceded by the much cheaper determination of the n-octanol water partition coefficient (certain chemicals, e.g. ionising ones, as described above are excluded).

It is, therefore, necessary to differentiate between properties
- which are a required input parameter for evaluating the behaviour of a chemical *(parts of evaluation)*,
- which represent a necessary piece of information which must be on hand before a special test can be performed properly *(pre-requisites)*, and
- such which are pieces of background information, necessary to optimize a certain test, or important for the interpretation of the test results *(guidance information)*.

For example, testing hydrolysis as a function of pH requires, besides the availability of a suitable analytical method, the knowledge of the water solubility as a pre-requisite, since the corresponding Test Guideline applies only to water-soluble compounds. The vapour pressure curve of the chemical under consideration is a piece of guidance information, because preference should be given to sealed or septum-closed tubes and head space avoided if the chemical is volatile. For the Modified MITI Tests (I) and (II) on ready and inherent biodegradability, again, an analytical method must be available for determining the concentration of the test material in the test solution and its empirical formula is required so that the

Table 4. Data components for the OECD minimum pre-marketing set of data

---

Chemical identification data

Name according to agreed international nomenclature, e.g. IUPAC
Other names
Structural formula
CAS-number
Spectra ("finger-print spectra" from purified and technical grade product)
Degree of purity of technical grade product
Known impurities, and their percentage by weight
Essential (for the purposes of marketing) additives and stabilisers and their percentage by weight

Production/Use/Disposal data

Estimated production, tons/year
Intended uses
Suggested disposal methods
Expected mode of transportation

Recommended precautions and emergency measures

Analytical methods

Physical/Chemical data

Melting point
Boiling point
Density
Vapour pressure
Water solubility
Partition coefficient
Hydrolysis[a]
Spectra
Adsorption – Desorption[a]
Dissociation constant
Particle size[a]

Acute toxicity data

Acute oral toxicity
Acute dermal toxicity
Acute inhalation toxicity
Skin irritation
Skin sensitisation
Eye irritation

Repeated dose toxicity data

14–28 days, repeated dose

Mutagenicity data

Ecotoxicity data

Fish LC 50 – at least 96 h exposure
Daphnia – reproduction 14 days
Alga   – growth inhibition 4 days

Degradation/Accumulation data

Biodegradation:
  screening phase biodegradability data (readily biodegradable)
Bioaccumulation:
  screening-phase bioaccumulation data (partitioning coefficient, n-octanol/water, fat solubility, water solubility, biodegradability)

---

[a] Only the screening part to be done for base set

theoretical oxygen demand (TOD) may be calculated. Information on the toxicity of the chemical may be useful guidance information for the interpretation of low results and in the selection of appropriate test concentrations.

The Expert Groups of the OECD Chemicals Testing Programme identified a test series which should enable, together with the necessary information on the identity, potential use patterns and marketing quantities, a first estimate of the possible exposure and effects behaviour of the substance to be made.

In 1982, as a result of these efforts, the OECD adopted the Decision C (82) 196 (Final) [27] which requires that in Member countries sufficient information be available on the properties of new chemicals before they are marketed to ensure that a meaningful assessment of hazard to man and the environment can be carried out. To implement this decision a *Minimum Pre-Marketing Set of Data (MPD)*, listed in Table 4, has been recommended that can serve as a basis for a meaningful first assessment of this kind.

Whether or not industry is required to submit all of the data in the MPD depends on the chemicals laws in the various countries. All of the nations of the European Communities have agreed to require an (almost identical) *Base Set* (Annex VII) of the Sixth Amendment of the *1967 EC Council Directive on the Classification, Packaging, and Labeling of Dangerous Substances* [28]. One advantage of requiring submission of MPD is that the assessor is assured of a generally adequate data set on the new chemical, and one that is consistent from chemical to chemical. The advantage for the manufacturer is that the data only have to be developed once since the MPD will be accepted by the other OECD countries for registration. There is, of course, a need for flexibility and all Test Guidelines [8] as well as the above-mentioned Decision C (82) 196 (Final) [27] contain a provision for substitution, addition, or omission of tests provided justification is supplied.

Application of MPD Elements for the Assessment of Major Environmental Pathways

In order to serve the needs of the assessors of new chemicals, preliminary environmental fate prediction has to be based solely on the minimum data provided at the time of premarketing notification. This means that the major environmental transport and transformation pathways sketched out in Fig. 1 either have to be expressible as a function of the related elements of this base set information or it must be possible to make reasonable assumptions on the basis of both *"typical cases"* and *"realistic worst cases"*.

Table 5 lists the major environmental pathways and contains a summary of selected mathematical representations of those processes which have to date been adequately studied. Only such correlations which can be expressed using MPD elements and are widely applicable to different chemical classes have been included. (More comprehensive compilations can be found in references [29–31].)

Obviously, for some pathways, adequate mathematical representations are not currently available, so that appropriate assumptions have to be made for screening purposes. Concerning the dispersion behaviour of a chemical compound within each environmental compartment, for example, experience has proved it reasonable to consider the different media taken collectively as a closed system, which reaches a *steady-state* condition after a finite period. In this case

Table 5. Environmental fate prediction – Major environmental pathways and their mathematical representations on the basis of MPD elements

Pathway	Available mathematical representation	Required MPD elements	Site specific parameters[a]	Source
Intramedial transport	Not applicable			
Air/water transfer and partitioning	$$k_{WA} = \dfrac{\dfrac{H}{RT} \times k_l \times k_g}{\dfrac{H}{RT} \times k_g + k_l} \times \dfrac{A}{V_W} \ [hr^{-1}]$$ $$k_{AW} = \dfrac{k_l \times k_g}{\dfrac{H}{RT} \times k_g + k_l} \times \dfrac{A}{V_A} \ [hr^{-1}]$$ $$K_{AW} = \dfrac{P \times M}{1000\,S \times RT}$$ Where k_{WA} = water/air transfer rate constant k_{AW} = air/water transfer rate constant K_{AW} = air/water distribution coefficient (applies only to equilibrium conditions) k_g = gas-phase exchange coefficient $[cm/s] = k_{gl} \times \sqrt{18/M}$ k_l = liquid-phase exchange coefficent $[cm/s] = k_{ll} \times \sqrt{44/M}$ k_{gl} = gas-phase exchange rate for water k_{ll} = liquid-phase exchange rate for CO_2 H = Henry's law constant $[Pa\text{-}m^3/mol]$ $= \dfrac{P \times M}{S\,1000}$ T = absolute temperature $[K]$ R = gas constant $(8.314\ Pa\text{-}m^3/mol\text{-}K)$	Molecular mass (M) [g/mol] Vapour pressure (P) [Pa] Water solubility (S) [kg/m³]	Interface area (A) [m²] Liquid volume (V_W) [m³] Air volume (V_A) Flow speed/turbulence	33 34
Air/soil transfer and partitioning	$$k_{SA} = \dfrac{\dfrac{H}{RTK_{D\varrho}} \times k_l \times k_g}{\dfrac{H}{RTK_{D\varrho}} \times k_g + k_l} \times \dfrac{A}{V_s}$$ $$k_{AS} = \dfrac{k_l \times k_g}{\dfrac{H}{RTK_{D\varrho}} \times k_g + k_l} \times \dfrac{A}{V_A}$$ $$K_{AS} = \dfrac{P \times M}{1000\,S \times RT \times K_{D\varrho}}$$ Where (symbols used already above not included) k_{SA} = soil/air transfer rate constant k_{AS} = air/soil transfer rate constant K_{AS} = air/soil distribution coefficient (applies only to equilibrium conditions)	Molecular mass (M) [g/mol] Vapour pressure (P) [Pa] Water solubility (S) [kg/m³] Adsorption coefficient (K_D) $\left[\dfrac{m^3\ water}{10^3\ kg\ sorbent}\right]$ $= K_{OC} \dfrac{\%\ org.\ carbon\ content}{100}$	Interface area (A) [m²] Soil volume (V_s) [m³] Air volume (V_A) [m³] Wind speed/turbulence Soil bulk density (ϱ) [kg/m³] % org. carbon content	35
Water/ sediment partitioning	$\log K_{OC} = 1.00 \log P_{OW} - 0.21$ While $$K_D = K_{OC} \dfrac{\%\ organic\ carbon\ content}{100}$$	n-Octanol water partition coefficient (P_{OW})	% organic carbon content	22
Leaching from soil to water	$$R_f = \dfrac{1}{1 + K_D (\frac{1}{\theta}^{2/3} - 1)\varrho}$$	Adsorption coefficient (K_D)	% organic carbon content pore fraction (θ) soil bulk density (ϱ)	19

Table 5 (continued)

Pathway	Available mathematical representation	Required MPD elements	Site specific parameters[a]	Source
Photolysis	$k_P = \phi \sum_\lambda \sigma_\lambda J'_\lambda \, [\mathrm{d}^{-1}]$ For screening purposes unit quantum yield ϕ [molec photons^{-1}] is assumed to set an upper limit for the rate constant of photolysis (k_P) so that the dominant process can be identified (cf. text)	UV-VIS Absorption spectra (cross section σ_λ [cm^2 molec^{-1}] at wave length λ)	$J'_\lambda =$ Solar intensities [photons cm^{-2} day^{-1}]	13 36 37
Oxidation HO	$k'_{OH} = \sum_i n_i \alpha_{Hi} \beta_{Hi} k_{Hi} + \sum_j \alpha_{Ej} k_{Ej} + \sum_l \alpha_{Al} k_{Al}$ Where $k_{OH}=$ first order rate constant for that process $\quad = k_{OH}[OH]$ k_{Hi}, k_{Ej}, k_{Al} are the reactivities of the ith hydrogen atom, jth carbon-carbon double bond, lth aromatic group, respectively. α_H and β_H account for the effect of substituents in α and β positions respectively	Structural formula	Concentration of OH	14
O$_3$	Does not at present exist. Rate constants for specific types of groups are determined from rate constants of generalized structures	Structural formula	Ozone concentration	14
ROO	Not available			
Hydrolysis	a) $1 \, \mathrm{d} \geqq t_{1/2} \geqq 1 \, \mathrm{y}$ b) $k_h = k_A[H^+] + k_B[OH^-] + k_N \; [s^{-1}]$ Where k_A and k_B are the second-order rate constants for the acid and base catalyzed processes, respectively, and k_N the pseudo-first order rate constant for neutral reaction	Hydrolysis as a function of pH – Preliminary test – Further testing beyond the preliminary test according to OECD Test Guideline 111	pH	38 39
Biodegradation by aerobic bacteria	$k > 0.07 \, \mathrm{d}^{-1}$ when "ready biodegradability" is observed $k = 0$ otherwise	Biodegradation: screening phase biodegradability data	Not considered	40 32
Biodegradation by anaerobic bacteria	Not available			
Bioaccumulation in water	$\log \mathrm{BCF} = 0.79 \log P_{OW} - 0.40$	n-Octanol water partition coefficient (P_{OW})	Not considered	41
Abiotic accumulation	$\log K_{OC} = 1.00 \log P_{OW} - 0.21$ $\log K_{OC} = 0.544 \log P_{OW} + 1.377$	n-Octanol water partition coefficient (P_{OW})	% organic carbon content	22 21

[a] At constant temperature

an even distribution of the substance within the media can often be presumed in first approximation.

The velocity with which steady-state is reached depends on the time the chemical needs for its transfers from one compartment to the other. Assumptions are at least in part necessary because the relevant theoretical basis for adequate calculation of all of these transfers at the MPD stage does not exist. So, the water/biota and water/sediment transfer processes are assumed to be so rapid that equilibrium partitioning between these phases prevails [32].

The biodegradation test results as provided in the MPD permit only a determination of whether the compound is readily biodegradable. For screening purposes, however, $0.07 \, d^{-1}$ can be used as a rough estimate of the biodegradation rate constant when "ready biodegradability" is observed. This corresponds to the minimum time allowed (10 days) for half of the material to be degraded once the acclimation has occurred. When the data are not available, the rate constants should be always assumed to be zero.

Similar assumptions and treatments apply to the hydrolysis rate.

Use of OECD Premarket Data In Environmental Exposure Analysis for New Chemicals

Once the Minimum Pre-Marketing Set of Data (MPD) had been proposed, it was thought important to develop principles to evaluate the hazard of chemicals before they are marketed. For this reason, in December 1979 the *OECD Hazard Assessment Project* was initiated.

The initial task of this project was to determine how information on the ultimate fate and effects of a chemical can be derived from the MPD, how these data can be combined to give an estimate of the hazard from exposure to that chemical, and to identify indications for relevant criteria for further testing and/or assessment. To accomplish this, three working parties were established.
○ *Health Effects* under the chairmanship of the United States
○ *Natural Environment Effects* under the chairmanship of Canada
○ *Exposure Analysis* under the chairmanship of the Federal Republic of Germany

The Exposure Analysis Working Party was given two tasks: to consider environmental partitioning, taking environmental transport and transformation into account, and to consider the exposure from all sources to humans and the environment.

For environmental exposure two expressions – *Potential Environmental Distribution (PED)* and *Potential Environmental Concentration (PEC)* – were considered [40], both of these making it possible to estimate exposure potentials of chemicals within environmental compartments of major concern.

The fundamental difference between the two approaches is that PEC gives an estimation of the non-equilibrium (point source) situation of a discharged chemical within distinct ecosystems whereas the PED procedure is an estimation of the widespread, equilibrium (or steady state) situation. The factors governing the latter type of exposure are largely those pertaining to the properties of the substance in the media and can normally be obtained from pre-marketing data elements.

In order to estimate localized concentrations (PEC), discharge patterns and the nature of specific receiving ecosystems must be defined. Also, the quantities of the chemical entering the environment at these sites must be known. A high degree of reliability and accuracy for hazard assessment from this analysis is usually not feasible without creation of environmental scenarios.

Of special interest among the methodologies for predicting PED is the model developed by Mackay [42–44] and adapted by Wood [45] which can be used at

Table 6. Environmental compartmentalization equations

Mass partitioning among the compartments	Equilibrium partitioning fractions
$$P_i = \dfrac{Z_i \times V_i}{\sum\limits_i Z_i \times V_i}$$	$$P_i = \dfrac{Z_i}{\sum\limits_i Z_i}$$

(i = air, water, sediment, soil, (aquatic) biota)
with the compartment specific fugacity capacities given by Mackay [42, 43]:

$$Z_{air} = 1/RT \qquad Z_{water} = 1/H \qquad Z_{soil/sediment} = K_D\varrho/H$$
$$Z_{biota} = P_{ow}B/H$$

where   R = universal gas constant (8.314 J/mol-K)
        H = Henry's law constant (Pa-m^3/mol)
        K_D = sorption coefficient (m^3 water/10^6 g sorbent)
        ϱ = sorbent density (g/cm^3)
        B = mass fraction of biota times lipid part
        P_{OW} = partition coefficient (n-octanol/water)

and V$_i$ the volumes of the different environmental compartments air, water, soil, sediment, and biota

---

several levels of sophistication depending on the chemical and environmental data available. The models are based on the concept of fugacity and assume equilibrium or steady-state conditions among a set of environmental compartments. In their simpler versions (no degradation and instantaneous equilibrium) they can be used for predicting *environmental compartmentalization* assuming that:

o the environment is a closed system consisting of air, water, sediment, soil, and (aquatic) biota compartments
o the chemical has reached steady state in the environment both with respect to interphase transfer and intraphase transport
o no degradation processes occur during the distribution.

Under these conditions a common fugacity prevails, so that, at the low concentrations which are relevant to environmental contaminants, the mass partitioning (P_i) among the compartments and the equilibrium partitioning fractions (P_i') can be calculated from the compartment-specific fugacity capacities (Z_i) as shown in Table 6.

Obviously, the essential physical/chemical data needed are

o molecular mass
o water solubility
o vapour pressure
o soil sorption constant ($K_{OC} = K_D$/fraction oc)
o partition coefficient (n-octanol/water)

Unlike the P_i values which give information on where most of the quantity of chemical may partition, the P_i' values reflect where the highest concentrations may occur. Both forms are therefore in use in assessments. For instance, the percentage of a chemical in biota is generally negligible from the point of view of mass balance but may be significant when considered as a relative concentration. Since the P_i values are dependent on the volumes V_i of the different environmental compartments under consideration, assumptions regarding compartment sizes

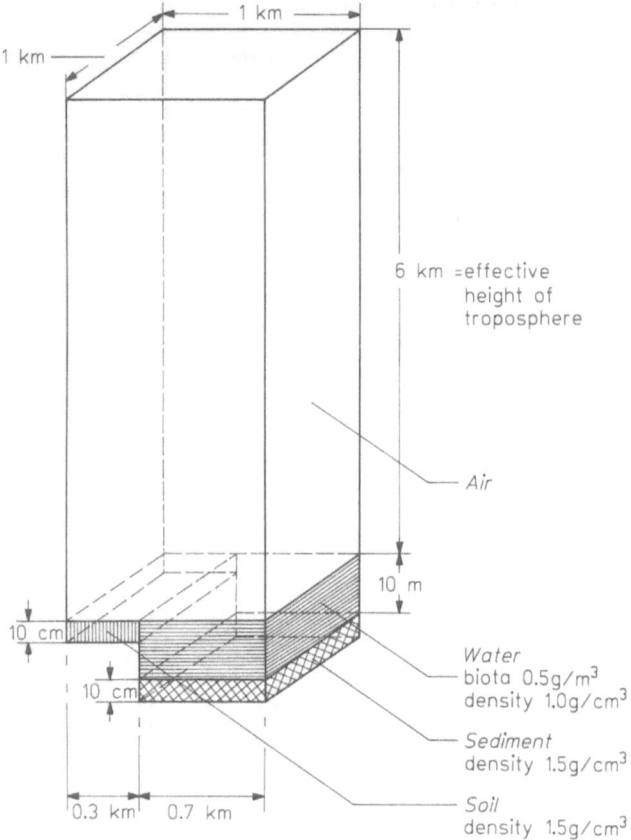

Fig. 4. Generic environment (Organic carbon content of soil: 2%, organic carbon content of sediment: 4%)

have to be made. The volumes that the Exposure Analysis Working Party selected as "standard" for future PED modeling efforts are shown in Fig. 4. A more detailed discussion of this generic environment can be found in reference [46].

Considering degradation and intercompartmental transfer processes, the simple versions of the models can be extended to provide a more realistic estimate of the Potential Environmental Distribution (PED) of a chemical. Taking into account the major environmental pathways and their mathematical representations listed in Table 5, the dynamic mass partitioning and dynamic concentration partitioning of a chemical among the main compartments air (A), water (W), and soil (S) under steady state conditions can be calculated as shown in Table 7.

Future Research Needs Regarding the Screening-Level Models

The environmental fate models described have not been systematically validated as yet, because multimedia data for test substances are not generally available. Obtaining these data is therefore of high priority.

Table 7. Dynamic distribution equations (steady-state conditions assumed)

Dynamic mass partitioning	Dynamic concentration partitioning

$$P_A = \frac{T_1}{T_1 + T_2T_3 + T_1T_4 + T_2T_5 + T_1T_6}$$

$$P'_A = \frac{T_1}{T_1 + (T_2T_3 + T_1T_4)V_A/V_W + (T_2T_5 + T_1T_6)V_A/V_S}$$

$$P_W = \frac{T_2T_3 + T_1T_4}{T_1 + T_2T_3 + T_1T_4 + T_2T_5 + T_1T_6}$$

$$P'_W = \frac{(T_2T_3 + T_1T_4)}{T_1V_W/V_A + (T_2T_3 + T_1T_4) + (T_2T_5 + T_1T_6)V_W/V_S}$$

$$P_S = \frac{T_2T_5 + T_1T_6}{T_1 + T_2T_3 + T_1T_4 + T_2T_5 + T_1T_6}$$

$$P'_S = \frac{(T_2T_5 + T_1T_6)}{T_1V_S/V_A + (T_2T_3 + T_1T_4)V_S/V_W + (T_2T_5 + T_1T_6)}$$

$$(A = air, \; W = water, \; S = soil)$$

Where the T terms stand for

$$T_1 = Q_A + [Q_W \times (k_{WA}/(k_W + k_{WA}))] + [Q_S \times (k_{SA}/(k_S + k_{SA}))]$$
$$T_2 = k_A + [(k_W \times k_{AW})/(k_W + k_{WA})] + [(k_S \times k_{AS})/(k_S + k_{SA})]$$
$$T_3 = Q_W/(k_W + k_{WA})$$
$$T_4 = k_{AW}/(k_W + k_{WA})$$
$$T_5 = Q_S/(k_S + k_{SA})$$
$$T_6 = k_{AS}/(k_S + k_{SA})$$

k_i = first order total degradation rate constant in the compartment i
k_{ij} = rate constant of transfer from compartment i to compartment j

Some data collected simultaneously on water, fish, sediment, and suspended solids by the Canadian Department of the Environment from Lake Superiour locations [47] show a rather fair to good agreement with appropriate predictions of equilibrium partitioning fractions using the simple fugacity approach [4]. This result was confirmed recently for a number of further chemicals monitored by the Japanese Environmental Agency so long as they do not degrade too rapidly within the receiving environmental setting, so that the partition equilibrium can be obtained [48].

For further support, microorganisms and model ecosystems might be useful surrogates of the real environment for validation purposes. As a start, a matching of the distribution patterns, which were established by experiments in an environmental standard system, and the patterns calculated by means of simple correlations could be shown by researchers of the German NATEC-Institute, Hamburg [49], but they still do not take into account a water compartment.

To further validate the models, their sensitivities to incremental variances in compartment size and to the accuracy of the input physical/chemical property and transformation data need to be determined. These will provide an estimate of the extent to which variations in model parameters and in the MPD set will affect the predicted results. If individual parameters are found to have a major effect, the accuracy and precision of test methods for these parameters must be reconsidered to ensure that useful results will be obtained.

Further Steps in the Exposure Assessment Process – More Sophisticated Approaches

At the premarketing stage, it is at present impossible to predict actual environmental concentrations of a chemical as a function of its probable spatial and tem-

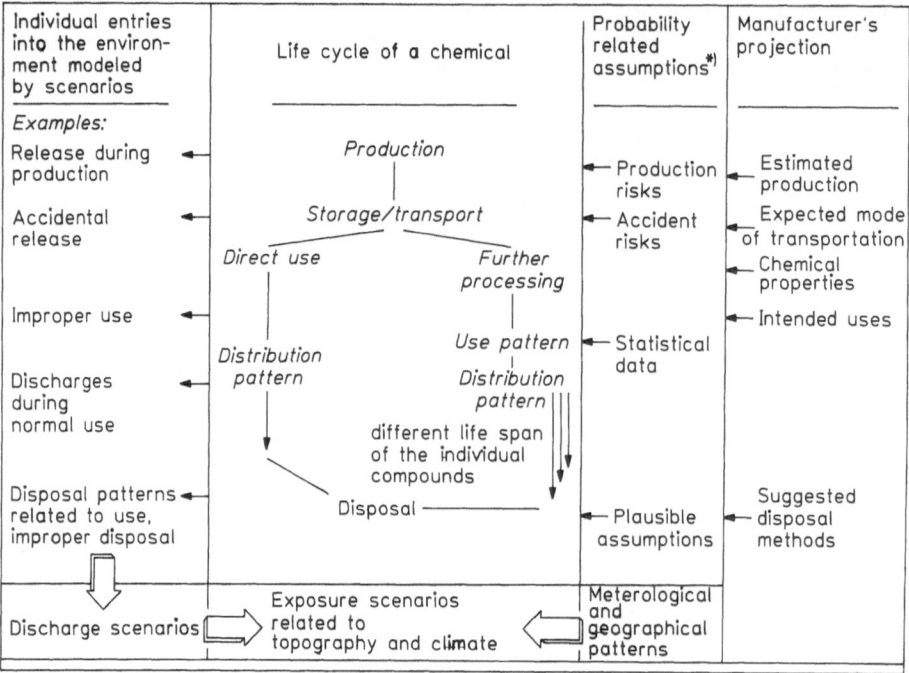

Fig. 5. Creation of environmental exposure scenarios on the basis of probability related assumptions (* taking into account statistics and experience with chemicals with similar structures)

poral distributions. This is mainly due to a lack of sufficient knowledge about the site-specific characteristics of the receiving environmental sectors. Unlike the distribution of persistent substances which reaches rather close global equilibrium conditions, especially the fate of less persistent compounds is highly dependent on the way of their entry into the environment [50, 51].

This entry may occur at any of the various stages of a chemical's economic life cycle and it should be feasible to interpret the discharge by an appropriate differential equations system. The formulation of the marginal conditions of such a system, however, would be quite complex and, therefore, strictly speaking not possible to generalize. For this reason, a set of transition functions has to be established in such a way that stochastical events can be used which involve a series of scenarios based on manufacturer's projections, what is known about chemicals with similar structures and further statistical data.

The structure of an appropriate system for creation of environmental exposure scenarios is illustrated in Fig. 5. The central part shows schematically the life cycle of a substance during which the individual entries of a chemical into the environment can be described by discharge scenarios. Together with the meteorological and geographical characteristics of a specific setting or region they may be combined to gain exposure scenarios related to the specific topography and climate of the area under consideration and which are indispensable for the prediction of environmental concentrations.

Conclusions

The first phase of the OECD Hazard Assessment Project is now finished. The basic "philosophy" developed may be written as an equation [52]:

Hazard = Potential for exposure × Potential to harm biological or other systems.

Although the Final Reports of the Working Parties do not contain a "recipe" for solving this equation, it is possible to derive a useful ranking system based on the principles of assessment developed in these documents and available data of a chemical at the pre-marketing stage [53].

The recommendation of MPD, including the provisions for its flexible application, was an important step towards the international acceptance of information about chemicals. This in itself, however, does not provide the international harmonization required.

Further details, therefore, are presently being worked out [27] with a view towards international agreements based on harmonized approaches. As a first step, *Provisional OECD Data Interpretation Guides (DIGs)* [54] were designed which may be consulted whenever initial hazard assessment of chemicals is contemplated. They present guiding principles for interpreting results from each of twenty-one MPD test parameters and three additional data elements being part of MPD. Their main purpose is to provide an international basis for sharing of experience; comments on their use will be collected, analysed and made available.

With respect to the screening-level models presented here, marginal conditions required can easily be harmonized. The more sophisticated approaches will have to be adopted by different countries to conditions related to their national topography and climate. It should be possible, however, to mutually agree on suitable scenarios for their statistical weighing which is a task of high priority.

References

1. OECD Council Recommendation C (74) 215: The Assessment of the Potential Environmental Effects of Chemicals, Paris, 14 Nov. 1974
2. OECD Council Recommendation C (77) 97 (Final): Guidelines in Respect to Procedures and Requirements for Anticipating the Effects of Chemicals on Man and in the Environment, Paris, 7 July 1977
3. OECD Chemicals Testing Programme, Step Systems Group, Final Report, Stockholm 1982, para 19
4. Klein, A.W., Schmidt-Bleek, F.: Significance and Limitations of Environmental Compartmentalization Models in the Control of New Chemicals Based on the OECD Minimum Pre-marketing Set of Data (MPD). In: Modeling the Fate of Chemicals in the Aquatic Environment (eds. Dickson, K.L. et al.) Ann Arbor Science Publishers, Inc., 1982, p. 73
5. Brown, S.M., Delos, C., Klein, A.W., Neely, B., Offutt, C., Wood, W.P., Yeh, G.T.: Mathematical Modeling of Chemical Fate. In: Modeling the Fate of Chemicals in the Aquatic Environment (eds. Dickson, K.L. et al.) Ann Arbor Science Publishers, Inc., 1982, p. 93
6. Klein, A.W., Haberland, W.: Evaluation of Current Methods for Predicting Environmental Fate of New Chemicals Based on European Directive 79/831/EEC, Annex VII. In: Chemicals in the Environment. Symposium Proceedings, 18–20 Oct. 1982, Lyngby-Copenhagen (eds. Christiansen, K. et al.), DIS Congress Service, Vanløse, Denmark, p. 98
7. Klein, A.W., Harnisch, M., Poremski, H.J., Schmidt-Bleek, F.: OECD Chemicals Testing Programme/Physico-Chemical Tests, Chemosphere *10* (2), 153 (1981)

8. OECD Guidelines for Testing of Chemicals (8000-TS-9781051) OECD Publications Office, Paris, 1981

9. Decision of the Council Concerning the Mutual Acceptance of Data in the Assessment of Chemicals C (81) 30 (Final). OECD, Paris, 12 May 1981

10. Schmidt-Bleek, F., Klein, A.W., Haberland, W.: Present Status of Hazard Assessment of Chemicals in the Environment – The Scientific View. In: Chemicals in the Environment. Symposium Proceedings, 18–20 Oct. 1982, Lyngby-Copenhagen (eds. Christiansen, K. et al.), DIS Congress Service, Vanløse, Denmark, p. 11

11. Davenport, J.E., Mill, T.H., Johnson, R.M., Chee-Liang Gu: Development and Evaluation of a Tiered Test System for Photochemical Degradability of New Chemicals in the Atmosphere Phase I. Final Report. Umweltbundesamt RED Contract No. 106-02-017, (1984)

12. Cupitt, L.T.: Fate of Toxic and Hazardous Materials in the Air Environment, Final Report EPA Contract No. 600/3-80-084

13. Mill, T.H., Mabey, W., Winterle, J., Davenport, J., Barich, V., Dulin, D., Lee, G., Moran, K., Tse, D.: Design and Validation of Screening and Detailed Methods for Environmental Processes, Final Report EPA Contract No. 68-01-6325

14. Hendry, D.G., Kenley, R.A.: Atmospheric Reaction of Organic Compounds. Final Report EPA Contract Nr. 68-01-5123

15. Soil Survey Staff and Collaborators: Soil Classification. A comprehensive system, 7th approximation. United States Department of Agriculture, Soil Conservation Service, Washington D.C. 1967

16. Isensee, A.R.: Laboratory Microecosystems. In: The Handbook of Environmental Chemistry. Hutzinger, O. (Ed.), Springer Verlag Berlin – Heidelberg – New York, Volume 2 Part A, 1980, 231

17. Gillet, J.W.: Terrestrial Microcosm Technology in Assessing Fate, Transport and Effects of Toxic Chemicals. In: Dynamics, Exposure, and Hazard Assessment of Toxic Chemicals. Haque, R. (Ed.). Ann Arbor, MI: Ann Arbor Science Publishers 1979, 231

18. Figge, K., Klahn, J.: Die Pflanzenstoffwechselbox, GIT Fachz. Lab. 26 (1982) 680–685

19. Hamaker, J.W.: The Interpretation of Soil Leaching Experiments. In: Environmental Dynamics of Pesticides. Haque, R., Freed, W.H. (Eds.). Plenum Press, New York and London, pp. 115–134, 1975

20. Briggs, G.G.: A simple relationship between the soil adsorption of organic chemicals and their octanol water partition coefficients. In: Proc. 7th Br. Insecticide Fungicide Conf., pp. 83–85, 1973

21. Kenaga, E.E., Goring, C.A.I.: Relationship between water solubility, soil sorption, octanol-water partitioning, and concentration of chemicals in biota. Aquatic Toxicology, ASTM STP 707, 78 (1980)

22. Karickhoff, S.W., Brown, D.S., Scott, T.A.: Sorption of hydrophobic pollutants on natural sediments. Water Res. *13*, 241 (1979)

23. OECD Chemicals Testing Programme. Expert Group Degradation/Accumulation Draft Final Report, Umweltbundesamt – Bundesrepublik Deutschland, Government of Japan, Berlin-Tokyo, Dec. 1979

24. Drescher-Kaden, V.: Nationale und internationale Forschungsaktivitäten und Ergebnisse auf dem Gebiet der Nutzung freilebender Tierarten als Indikatoren für die Belastung der Umwelt – insbesondere des Menschen – durch Umweltchemikalien. Expertise on behalf of the Federal Minister for Youth, Family and Health (Ed.), Bonn, Dec. 1976

25. Vorkommen von Organohalogenrückständen in freilebenden Tierspezies. Research Report No. 037110 on behalf of the Federal Minister for Research and Technology, Bonn, Dec. 1978; Kernforschungsanlage Jülich, Special Reports, No. 45, July 1979 (ISSN 0343-7639)

26. Veith, G.D., Macek, K.J., Petrocelli, S.R., Carroll, J.: An evaluation of using partition coefficients and water solubility to estimate bioconcentration factors for organic chemicals in fish. Aquatic Toxicology, ASTM STP 707, 116 (1980)

27. Decision of the Council Concerning the Minimum Pre-Marketing Set of Data in the Assessment of Chemicals [C (82) 196 (Final)]. OECD, Paris, 8 Dec. 1982

28. European Communities' Council Directive 79/831/EEC of 18 September 1979 amending for the 6th time the Council Directive 67/548/EEC of 27 June 1967 on the Approximation of the Laws, Regulations, and Administrative Provisions Relating to the Classification,

Packaging and Labeling of Dangerous Substances. Off. J. Eur. Comm. Council L 259. October 15, 1979, p. 10

29. Roberts, J.R. et al.: A Screen for the Relative Persistence of Lipophilic Organic Chemicals in Aquatic Ecosystems – An Analysis of the Role of a Simple Computer Model in Screening. National Research Council Canada, No. NRCC 18570, 1981

30. Lyman, W.J., Reehl, W.F., Rosenblatt, D.H. (ed.): Handbook of Chemical Property Estimation Methods – Environmental Behaviour of Organic Compounds. McGraw – Hill Book Company, 1982

31. Fong, C.V., Clermann, R.J., Hushon, J.M.: Hazard Evaluation of New Chemicals. Final Report. Umweltbundesamt Contract No. 10701009. October 1979

32. Hushon, J.M., Klein, A.W., Strachan, W.M.J., Schmidt-Bleek, F.: Use of OECD Premarket Data in Environmental Exposure Analysis for New Chemicals. Chemosphere 12 (6), 878 (1983)

33. Liss, P.S., Slater, P.G.: Flux of Gases Across the Air-Sea Interphase. Nature 247, 181 (1974)

34. Mackay, D., Leinonen, P.J.: Rate of Evaporation of Low Solubility Contaminants from Water Bodies to Atmosphere. Environ. Sci. Technol. 9, 1178–80 (1975)

35. OECD Working Party on Exposure Analysis. Room-Document Expo-81.27/D, USA, CDN: Report on the Workshop on Environmental Compartmentalization. Washington, March 28, 1981

36. Schere, K.L., Demerjian, K.L.: Calculation of Selected Photolytic Rate Constants over a Diurnal Range. Final Report EPA 600/4-77-015

37. Zepp, R.G., Cline, D.M.: Rates of Direct Photolysis in Aquatic Environments. Environ. Sci. Technol. 11 (3), 359 (1977)

38. Smith, J.H., Mabey, W.R., Bohonus, N., Holt, B.R., Lee, S.S., Chou, T.W., Venberger, D.C., Mill, T.H.: Environmental Pathways of Selected Chemicals in Freshwater Systems. Part I and II, Final Report, EPA 600/7-78-074 (1978)

39. Mill, T.H., Mabey, W., Bomberger, D.C., Chou, T.W., Hendry, D.C., Smith, H.: Laboratory Protocol for Evaluating the Fate of Organic Chemicals in Air and Water. Final Report, EPA Contract No. 68-03-2227 (1981)

40. OECD Working Party on Exposure Analysis, Final Report. Umweltbundesamt Berlin, March 1982

41. Veith, G.: Private Communication to W.P. Wood, from Environmental Research Laboratory, U.S. EPA, Duhet Minn., elaborating on the material in J. Fish. Res. Board Can. 36, 1040 (1979)

42. Mackay, D.: Finding Fugacity Feasible. Environ. Sci. Technol. 13 (10), 1218 (1979)

43. Mackay, D., Paterson, S.: Calculation of Environmental Partitioning and Persistence of Chemicals Using the Fugacity Approach. Environ. Sci. Technol. 15 (9), 1006 (1981)

44. Mackay, D., Paterson, S.: Fugacity revisited. Environ. Sci. Technol. 16 (12), 654 A (1982)

45. Wood, W.P.: Comparison of Environmental Compartmentalization Approaches. OECD Working Party on Exposure Analysis. Room-Document Expo – 80.21/USA (1981)

46. Neely, W.B., Mackay, D.: An Evaluative Model for Estimating Environmental Fate. In: Modeling the Fate of Chemicals in the Aquatic Environment (eds. Dickson, K.L. et al.) Ann Arbor Science Publishers, Inc., 1982, p. 127

47. Strachan, W.M.J.: Personal communication

48. Yoshida, K., Shigeoka, T., Yamauchi, F.: Non-Steady-State Equilibrium Model for the Preliminary Prediction of the Fate of Chemicals in the Environment. Ecotoxicol. Environ. Saf. 7, 179 (1983)

49. Figge, K., Klahn, J., Koch, J.: Testing of Chemicals by Evaluation of Their Distribution and Degradation Patterns in an Environmental Standard System. International Symposium: Testing in Ecology-Methods and their Evaluation, Munich-Neuherberg, Mai 17–19, 1982

50. Frische, R., Klöpffer, W., Schönborn, W.: Bewertung von organisch-chemischen Stoffen und Produkten in Bezug auf ihr Umweltverhalten – chemische, biologische und wirtschaftliche Aspekte. Final Report. Umweltbundesamt Contract No. 101 04 009/03 (1979)

51. Klöpffer, W., Rippen, G., Frische, F.: Physicochemical Properties as Useful Tools for Predicting the Environmental Fate of Organic Chemicals. Exotoxicol. Environ. Saf. 6, 294 (1982)

52. Klöpffer, W.: Assessment of Environmental Hazard of Substances and Mixtures – Integration of the Final Reports of the three Working Parties within the OECD Hazard Assessment Project. Umweltbundesamt Contract No. 106 04 014 (1983)
53. Klein, A.W., Haberland, W.: Environmental Hazard Ranking of New Chemicals based on European Directive 79/831/EEC, Annex VII. In: Chemicals in the Environment. Symposium Proceedings, 18–20 Oct. 1982, Lyngby-Copenhagen (eds. Christiansen, K. et al.), DIS Congress Service, Vanløse, Denmark, p. 419
54. DIGs – Data Interpretation Guides for Initial Hazard Assessment of Chemicals – Provisional. OECD, Paris 1984

Biodegradation and Transformation of Recalcitrant Compounds

A. H. Neilson, A.-S. Allard, M. Remberger

Swedish Environmental Research Institute
Box 21060, S-100 31 Stockholm, Sweden

Introduction

The persistence of man-made chemicals may be a necessary requirement for compounds designed specifically for technical use, or of compounds, like insecticides directed towards specific targets. In general, however, persistence of man-made chemicals in the natural environment is undesirable. Studies on degradation including both biotic and abiotic processes therefore occupy a central position in environmental hazard assessments. Whereas a more general view has been presented by Landner [1], this review deals exclusively with problems related to studies on biodegradability.

Substances which persist in the environment for extended periods of time in all of the environments examined have been termed recalcitrant [2]. This review is devoted to such compounds. However, for reasons which will become clear, the term "recalcitrant" is relative rather than absolute. Nonetheless, it should perhaps be pointed out that recalcitrant compounds may not necessarily be synthetic. Natural substances like antibiotics, or halogenated metabolites produced by marine algae and animals [3] may also be persistent. Their biodegradability, however, is unknown and indeed has seldom been investigated.

In this review, the apparently synonymous terms transformation and degradation are used. The distinction between them is useful and it is therefore necessary to provide pragmatic definitions. Transformation is used for situations in which the major structure of the substrate remains unaltered, and only one or two reactions such as oxidation, reduction, hydrolysis or methylation has taken place. Typical examples of transformations are the hydroxylation of steroids, the epoxidation of aromatic hydrocarbons by fungi and the hydrolysis of aromatic nitriles without cleavage of the aromatic ring. On the other hand, conversion of phenol to muconolactone or of naphthalene into salicylate would be considered degradations.

The transformation and complete degradation of readily degraded compounds to carbon dioxide (mineralization) have been extensively investigated, and standardized test procedures have been developed [4]. The situation for less readily degraded compounds is, however, considerably more complicated. Low degradation rates make the application of conventional procedures using, for example, measurements of growth rates, oxygen consumption, evolution of carbon dioxide or decrease in the concentration of dissolved organic carbon, impractical or even impossible. Alternative methods must therefore be sought.

The procedures used in this laboratory form the basis of this review and incorporate two basic principles, namely:

1. The use of pure cultures of microorganisms isolated after selective enrichment.
2. Incubation of dense cell suspensions and periodic sampling for analysis of the substrate and putative metabolites.

An important consideration is the choice of test organism. Attempts to develop a standardized test organism, or mixture of organisms, are beset with difficulties and may even be misdirected. The crucial question is not whether a given compound *may* be degraded but whether, in a natural system, it *is* degraded.

An instructive case is that of nitrilotriacetate. Although this compound has been shown to be degradable in a variety of situations, attempts to isolate estuarine bacteria able to degrade this were unsuccessful even though the relevant organisms could be isolated from incoming streams [5]. This example supports the view advocated in this review, that the most relevant test organisms are those isolated from a natural system into which the compound has been discharged.

Therefore, it was decided at the outset, that test organisms should be isolated from natural samples and that strains from culture collections should not be used. Artificially constructed strains [6–10], were likewise not considered as suitable for the present purpose. Because such organisms would not be expected to occur naturally, use of them in degradation studies might provide a misleading view of the degradability of the compound being examined. It should, however, be emphasized that investigations using such strains, and those probing the evolution of catabolic pathways in bacteria [11, 12] have provided valuable insights into the genetics and biochemistry of degradation. However, their significance in natural situations is hitherto unresolved.

Furthermore, attention should be drawn to the possibility of using immobilized cells. Various procedures have been developed and are summarized in review articles [e.g. 13]. The use of such systems is attractive from the following points of view:

1. Activity is stable over long periods of time and is highly reproducible.
2. Isolation and identification of metabolites is particularly easy, because substantial quantities are potentially available.
3. It is possible to carry out experiments with mixtures of immobilized cells under highly defined conditions.

Although such systems are not representative of natural situations, and there are technical problems in maintaining sterility, the possibility of their application to studies in biodegradation should be kept in mind.

The biodegradability of a compound is a function of many parameters other than structure. These parameters include the potential of the compound to associate with both organic and inorganic material in the environment, as well as physico-chemical factors such as pH, temperature, oxygen tension and salinity. A fairly extensive discussion has therefore been devoted to these and other relevant determinants.

Moreover, the views presented here are inevitably influenced by experience with the procedures used in this laboratory. An attempt has been made, however, to adopt a rather wide perspective and to evolve principles generally applicable to studies of biodegradation. Nevertheless, a somewhat formal mechanistic approach has been adopted and is justified on the one hand by its success in serving as a basis for delineating broad principles, and on the other hand, by providing insight into details of the relevant metabolic processes.

The references to the voluminous literature are eclectic rather than comprehensive. Apology is therefore offered to those whose studies appear to have been overlooked, as well as to the hopefully fewer, whose work has been misinterpreted. In the preface to a recent book [14], Postgate wrote "Only occasionally have I been concerned with priorities; I have taken the view that, for a given detail, the most useful citation is usually to recent papers, or even reviews: the reader may then back-track from these if he wishes to go into the matter in depth." These remarks apply equally to this review. Attention is therefore directed to a number of review articles [15–26], the proceedings of a symposium [27] and to some books [8, 28–30] which contain articles pertinent to the subject of this review.

Extensive changes in nomenclature over the years, together with the adoption of more sophisticated methods in classification, have resulted in some degree of uncertainty concerning organisms used by earlier workers. Where the currently accepted name is known to the authors, this has been given in parenthesis. It is hoped this may be of some value in assessing the metabolic potential of the same (or different) organisms.

Experimental Procedures

The Organisms

In all of our own studies and in the majority of those discussed in this review, pure monocultures were employed. Although it is obvious that these are very seldom encountered in natural situations, their use does, however, offer significant advantages. Pure monocultures can be obtained by enrichment from natural samples (water, sediment, soil). Because they represent highly reproducible biological material, detailed investigations can readily be carried out over extensive periods of time, and comparisons can be made with literature data for other organisms. It is therefore desirable that fairly detailed characterization of the organisms be carried out even though precise taxonomic location within existing schemes may not be possible. In a review dealing with flavobacteria [31], it has aptly been pointed out that clinical strains subjected to a relatively uniform and selective environment may be much more homogeneous that those strains iso-

lated from the environment which are subjected to a wide range of potentially selective pressures. The role of co-cultures and of syntrophy is discussed later.

Enrichment

Because enrichment plays a central role in the test system advocated, a somewhat
detailed discussion is presented. This discussion attempts to provide a general
background and also points out certain experimental difficulties.

A great deal of the knowledge and expertise in this area belongs to the folklore
of microbiology. Reference should be made to van Niel's Marjory Stephenson
Memorial Lecture [32] for many illuminating comments and interesting examples
of enrichment as well as for key references to the classical literature. Attention
is also drawn to a valuable article [33] which summarizes enrichment procedures
for non-fluorescent pseudomonads, and which provides examples illustrating
principles relevant to all enrichment cultures.

It seems specially fitting to begin by quoting from van Niel who was in a
unique position to evaluate some fifty years of activity in this area of microbiology.

"It was especially the work of Winogradsky on a restricted scale [34] and of Beijerinck [35]
on a far more general basis, that has made the microbiologist familiar with the possibility of using
a direct ecological approach to microbial ecology, by means of the elective, selective, or enrichment culture technique."

The following should be particularly emphasised:
1. The isolation of an organism with a given metabolic activity from a sample
   clearly substantiates the existence of microbes with the desired metabolic capability. Moreover, it seems justifiable to use this as evidence in support of the
   environmental significance of studies using such organisms. To quote van Niel
   once again:
   "But once an elective culture method for a particular microbe is available, it may be safely
   concluded that this organism will also be found in nature under conditions corresponding in
   detail to those of the culture, and that it will carry out the same transformations."
2. The method has been extensively and successfully utilized for isolating organisms with unusual catabolic activity.
3. An almost unrestricted range of experimental conditions may be used which
   allows the isolation of organisms from virtually all habitats.

The following constraints, for example, may easily be applied: low and high
temperatures (psychrophiles and thermophiles), low and high pH values (acidophiles and alkalophiles [36], low substrate concentrations (oligotrophs [37, 38]),
high NaCl concentration (halophiles), hyperbaric pressure (barophiles [39–41]),
anaerobic conditions (anaerobes). With the exception of the last, few of these
have been employed in practice. The fact that they could provide strains relevant
to almost any natural niche seems, however, to be a highly attractive feature of
this methodology.

Enrichments using substances having adequate water solubility present no
problems unless they are toxic. Volatile compounds such as toluene [42] or hydrogen cyanide (43) may be supplied in the gas phase. If the substance is insoluble
in water, it is generally supplied as a suspension (cellulose, aromatic hydrocar-

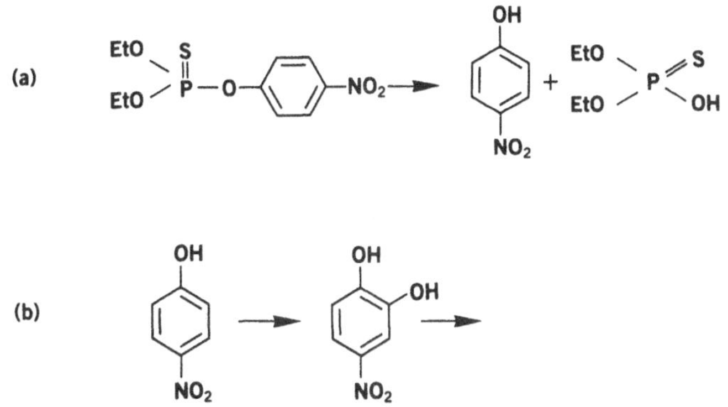

Fig. 1 a,b. Metabolism of Parathion by a consortium consisting of (**a**) *Pseudomonas stutzeri* and (**b**) *Pseudomonas aeruginosa* [44]

bons). Thus, there are essentially no problems associated with insoluble substrates so far as the enrichment procedure itself is concerned.

Enrichment may be carried out by two significantly different experimental procedures: 1. in continuous and 2. in batch culture. The first of these is an attractive procedure from several points of view and is particularly suitable for the isolation of truly oligotrophic organisms [37, 38], because under steady-state conditions, extremely low substrate concentrations are maintained. This method has also been used successfully for isolating mixed cultures where the product from the first organism is toxic to the second organism which is responsible for the ultimate degradation of the intermediate metabolite (Fig. 1). Furthermore, this procedure is particularly appropriate for application to substances with known (or suspected) toxicity to microorganisms. Because experiments in continuous culture may be carried out over prolonged periods of time, this procedure is eminently suitable for the study of stable consortia which have been established [45–47]. Substrate concentrations comparable to those which might occur naturally may also be imposed by the appropriate nutrient limitation. Continuous enrichment from soil or sediment samples may also be accomplished by percolation. If the substrate is insoluble, it may be mixed in the solid phase used as a source of inoculum.

For non-continuous enrichment, two different procedures may be employed. In one, the partly dried sample (soil or sediment) is sprinkled onto the surface of solid medium containing the substance being examined: after incubation, growth surrounding the solid particles may be removed and streaked onto fresh medium until a pure culture is obtained. Alternatively, enrichment may be carried out in liquid medium containing the substrate of interest. Successive transfer may then be made into fresh medium before isolation of the desired microorganisms is attempted. Workers differ widely in the number of transfers carried out. Some use only one or two while others employ between five and ten. Because we have usually been interested in relatively recalcitrant substances, we have generally employed four.

Table 1. Examples of syntrophic associations in which one organism provides a vitamin necessary for the growth of the other

Vitamin	Donor	Receptor	Reference
Cyanocobalamin	*Streptomyces* sp.	(a)	[50]
Biotin	*Ps. oleovorans*	*Mycotorula* sp.	[51]
	Methylocystis sp.	*Xanthobacter* sp.	[52]
Thiamin	*Ps. aeruginosa*	*Pseudomonas* sp.	[53]

(a) Unidentified bacterium

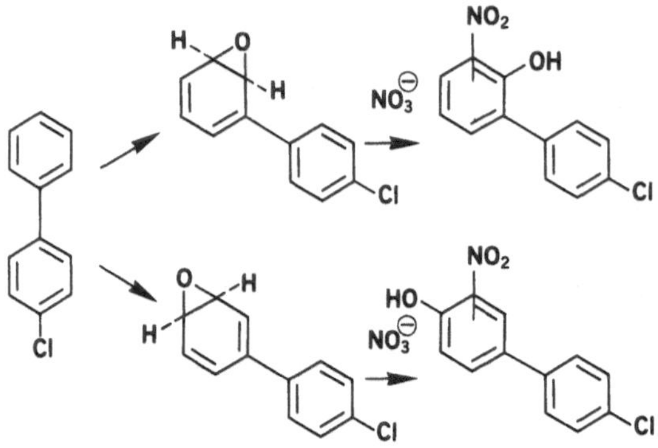

Fig. 2. Transformation of 4-chlorobiphenyl with synthesis of nitro metabolites from an intermediate epoxide [56]

Whichever procedure is adopted, a mineral base medium must be chosen. Although this is generally not critical, that of Palleroni & Doudoroff [48] or the modified medium of Van Veen [49] is suggested. It has become increasingly clear that organisms having an absolute requirement for vitamins are frequently encountered and that this is, at least in some cases, the reason for the existence of syntrophic associations. Some examples are given in Table 1. Therefore the addition of a mixture of the common vitamins (thiamin, biotin, pantothenate, and B_{12}) is routinely made in this laboratory.

The choice of nitrogen and sulphur source must also be considered. Most microorganisms (with the notable exception of those involved in sulphur metabolism) are able to utilize sulphate as a source of sulphur. Nitrogen requirements are more variable and may be fulfilled by nitrate, ammonium or molecular nitrogen. Nitrate may also serve as electron acceptor under anaerobic conditions [54]. However, certain complications may arise when nitrate is used [55] because of its possible incorporation into metabolites. This may be illustrated by the following examples:

1. The metabolism of 4-chlorobiphenyl in the presence of nitrate whereby nitrite produced is incorporated into the metabolites (Fig. 2).
2. In the presence of nitrate, chlorinated anilines may undergo a variety of transformations which may plausibly be rationalized on the basis of the intermedi-

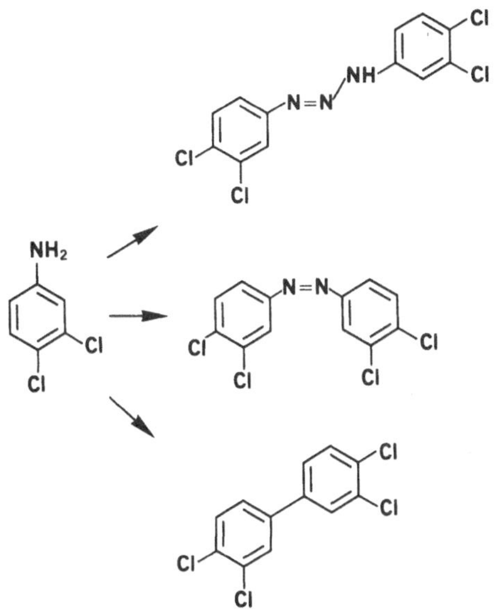

Fig. 3. Transformations of 3,4-dichloroaniline in presence of nitrate; products may plausibly involve diazonium intermediates [57, 58]

Table 2. Examples of substrates serving as multiple sources of carbon, nitrogen, sulphur or phosphorus

Source of	Substrate	Reference
Carbon + nitrogen	Nitriles	[60]
Carbon + sulphur	4-methylbenzenesulphonate	[61]
	Dimethyl sulphoxide	[62]
Nitrogen + sulphur	Thiocyanate	[63)
Carbon + nitrogen + sulphur	2-aminoethane sulphonate	[64]
Carbon + nitrogen + phosphorus	2-aminoethyl phosphonate	[65]

ate formation of a diazonium cation (Fig. 3). Such transformations are quite distinct, however, from peroxidase-mediated reactions [59].

If, however, the substrate contains nitrogen, sulphur or phosphorus, and organisms are sought which utilize the substrate as more than a source of carbon, inorganic additions of the relevant element must clearly be omitted. Some examples of substrates serving as sources of more than carbon are given in Table 2.

Enrichments may readily be carried out under virtually any conditions of pH, salinity, temperature, and oxygen tension. For anaerobic enrichments, the degree of anaerobiosis required depends critically on the nature of the organisms sought. This, together with the greater nutritional demands of certain groups (e.g. clostridia), must be carefully kept in mind.

For batch enrichments in liquid media, the size of the inoculum used for transfer may be critical. Although, the requirement of anaerobes for CO_2 is well established, it has become clear, that a number of other heterotrophic bacteria, for example, *Escherichia coli* [66], methylotrophs [67] and several organisms primarily of clinical interest, may require supplementation with carbon dioxide. Although CO_2 is normally produced in adequate amounts by respiration of the substrate, if rates of respiration are low or if only small inocula are used, carbon dioxide limitation may result in extended lag phases.

It should be noted that, at least in our experience with less readily degraded substrates, enrichment frequently results in a population of organisms of which only a fraction may be able to degrade the initial substrates. This problem is discussed later.

Isolation

Isolation of the relevant organisms may present problems of varying severity. If the substance is sufficiently soluble in water, plates of mineral medium containing the substrate as sole source of carbon (or nitrogen, or sulphur, or phosphorus) may be prepared, and a loopful of the enrichment culture can be spread on the surface. After incubation under the desired conditions, isolation, and purification then follow established procedures. In this laboratory, pure strains are preserved at $-70\,°C$ immediately after isolation. Thereby, possible alterations due to successive transfer on non-selective media are avoided.

While water-insoluble substances present few difficulties in the enrichment step, isolation of the relevant organisms may be a much more serious problem. A number of devices have been used, though no universally applicable procedure has emerged:
1. The substrate may be supplied as a suspension suitably dispersed in the agar medium (e.g. cellulose, paraffin).
2. The substrate may be adsorbed on silica powder which is then dispersed as above [68].
3. The substrate is supplied as a film on the surface of agar plates by spraying a solution of the compound dissolved, for example in acetone or ether: growth is then assessed from the formation of clear zones surrounding the colonies [69–71].
4. The substrate may be supplied in the vapour phase. This procedure has effectively been employed for toxic substrate of adequate volatility (e.g. toluene [42], hydrogen cyanide [43].
5. On the basis of plausible metabolic pathways, an alternative substrate may be chosen which can be incorporated into an agar medium at concentrations sufficient to produce visible growth. A serious pitfall in the application of this procedure is that some organisms are unable to utilize compounds for growth even when these are established metabolic intermediates. Presumably the cells are simply impermeable to the substrate. Clearly established examples are the inability of *Pseudomonas saccharophila* to utilize exogenous glucose for growth

[72] and the failure of wild-type strains of fluorescent pseudomonads which are able to degrade aromatic compounds via *cis,cis* muconate, to use this as a substrate for growth. Indeed, use of exogenous *cis, cis* muconate in enrichments leads to the isolation of strains of *Pseudomonas acidovorans* which, though having a functional β-ketoadipate pathway, are unable to degrade aromatic compounds [73]. Some alkane-utilizing bacteria are able to utilize only alkanes as substrates and are totally unable to use of corresponding carboxylic acids presumably produced during metabolism of the alkanes [74]. A comparable situation was encountered during enrichments with 1- and 2-naphthoates. Here organisms were readily isolated on solid media containing the appropriate substrate but were unable to make use of a wide range of seemingly plausible metabolites. Thus caution must be exercised against the uncritical application of this otherwise attractive procedure.

6. Complex media such as nutrient agar, soy tryptone agar etc. are frequently used. This is, however, a somewhat dangerous procedure for two reasons:
   (i) the desired organism may be unable to grow on rich peptone media (e.g. obligate methylotrophs, alkane oxidizers)
   (ii) if the inoculum contains several organisms, the competitive advantage of the organism desired may be lost during transfer from a selective to a nonselective medium. Overgrowth of other metabolically uninteresting organisms may then readily occur. Furthermore, organisms are frequently isolated which are unable to utilize either the initial substrate or metabolites produced from it.

Finally, attention must be drawn to some particularly troublesome problems for which satisfactory solutions may not always be found. Difficulty may be encountered during attempted isolation of filamentous organisms by overgrowth of "unwanted" organisms. For actinomycetes, a number of methods have been used which are based upon selective media, exposure to elevated temperatures or addition of antibiotics [refs. in 75]. A number of "mechanical" devices have also been advocated. It has been suggested that growth of actinomycetes may be encouraged at the expense of other organisms by filtering the sample through membrane filters and inverting these before incubation. Alternatively, the filter may be removed from the agar surface after primary incubation. Further incubation then allows the development of mycelium which has penetrated the filter [76]. Additionally, the use of high concentrations of salt has been suggested [77] for isolating species of *Streptomyces*.

Since slow-growing organisms such as mycobacteria present special problems, various decontamination procedures have been suggested generally involving treatment with alkali and/or hypochlorite [78]. In view of the metabolic versatility of representatives of the above groups and the scant information on their distribution and significance in natural environments, investigations taking advantage of these procedures are justifiable.

Use of serial dilution has also been used in certain circumstances, although this is a time-consuming method. It has been successfully used, however, for slow-growing organisms occurring in only low numbers (e.g. bacteria degrading polyvinyl alcohol [79] and for organisms which are unable to produce colonies on agar media (e.g. *Thermomicrobium fosteri* [80]).

Source of the Inoculum

In many studies on bio-degradation, mixed cultures from communal sewage-treatment plants are used as the source of the organisms. Apart from the dubious importance of such organisms in natural situations, there seems to be an important objection to their use. Such organisms will have been exposed to a milieu containing readily degradable substrates. Therefore, unless the enzymes required for degradation of the compound under examination are constitutive, such exposure may be counter-selective to organisms able to degrade unusual or recalcitrant compounds. For both of these reasons, use was not made of such sources in our studies. Instead, inocula were taken from soils or sediment samples which, at least in some cases, had been subjected to previous exposure to the substances. It was hoped that some measure of ecological relevance would thereby be introduced.

Design of Metabolic Experiments

It has been pointed out that, for less readily degraded compounds, rates of oxygen uptake, CO_2 evolution and growth rates of organisms may be extremely low. These are not then suitable parameters, unless radioactively labelled substrates are available. Although an extensive range of these substrates is commercially available, it is clear that in many instances the synthesis of the compound must be undertaken by the laboratory carrying out the investigation. Our interest was directed primarily towards a range of chlorinated guaiacols and related compounds. The synthesis of ring-labelled compounds would itself have been a major undertaking and was not considered realistic. Unfortunately, this must be a common situation with complex structures. Alternative procedures must then be developed.

For substrates containing chlorine, nitro, or sulphonate substituents, it may be convenient to analyse for production of inorganic chloride, nitrite or sulphite.

Most studies in this laboratory, however, have relied on analysis of metabolites produced during incubation of the substrates with dense cell suspensions of organisms isolated by enrichment of natural samples. Some workers have sought to prevent cell proliferation by addition, for example, of chloramphenicol. Because of possible complications and the lengthy periods of incubation often necessitated in our studies, this was not carried out.

Dense cell suspensions were incubated with the substrates under relevant conditions and samples were removed periodically for analysis of the concentration both of the substrate and of putative metabolites produced from it. It was of course appreciated, that such conditions will seldom or never prevail under natural circumstances. The use of such experimental procedures appears, however, to be sanctioned by their widespread use in respirometric studies of biodegradability as well as by their practical convenience. Nevertheless, some other limitations are obvious; for example, under natural conditions an organism will seldom be exposed to only one substrate. Two aspects are particularly significant:

1. The role of the growth substrate in inducing the enzymatic machinery required for metabolism of a structurally disparate substance.

2. The effect of another readily degraded substrate present during metabolism of the test substrate.

In order to avoid further confusion, we have termed the latter case "concurrent metabolism" so as to avoid conflicting usage of the term "co-metabolism." These aspects are discussed fully in a subsequent section. Details of the experimental procedure used by us are given elsewhere [81].

Outline of Experimental Methods

Procedures used in this laboratory for growth of cells, preparation of cell suspensions and incubation have been described [82], so that they need not be repeated here. Emphasis is rather placed on aspects not covered in publications and particularly on areas of experimental difficulty.

Enrichment is generally carried out by batch procedures using a mineral-base medium modified [82] from that of Van Veen [49]. Biotin, thiamine, pantothenate, and vitamin B_{12} are routinely added, and ammonium is the nitrogen source. The inoculum is a sample of water, or sediment or soil from the area in which the persistence of the substrate is to be assessed. Transfers are generally made 4–5 times but, from our experience, a consortium of organisms is still frequently present.

Soluble substrates are added to the medium at concentrations of 500 mg.L^{-1} which will generally yield visible growth if the substrate is degraded. Such concentrations are clearly too high for toxic substrates such as chloroanisoles and nitrophenols, so that considerably lower concentrations (50 mg.L^{-1}) may be necessitated.

Insoluble substrates are supplied in amounts greatly exceeding their solubility in water. However, the actual concentration in the medium is lower so that problems with toxicity are minimal. Volatile substrates such as methane are supplied in the gas phase and incubation carried out in serum bottles sealed with crimp caps.

Anaerobic enrichments are carried out by incubating medium containing the test substrate, pre-reduced with thioglycollate and cysteine in anaerobe jars. Yeast extract (200 mg.L^{-1}) is routinely added since many anaerobes of environmental significance (e.g. clostridia) are unable to grow in unsupplemented medium. Such conditions naturally preclude isolation of the more nutritionally fastidious organisms as well as strict anaerobes such as methanogens.

The experimental difficulty in isolating pure cultures of anaerobes does not need to be repeated, and the absolute necessity of using pure cultures (or cultures of known components) cannot be too strongly emphasized.

Nonetheless, universally applicable methods for isolation of the appropriate .organisms cannot be prescribed. Ideally, the enrichment substrate should be used to prepare plates of solid medium, but this may be impossible because of limited solubility. Pitfalls resulting from the use of complex media and analogue substrates have already been discussed.

Incubations were carried out under a variety of circumstances depending on the solubility and volatility of the test substrate. Volumes used were determined by the size of sample required for analysis and the duration of the experiments.

For open systems, it was essential to take into account evaporation losses, and this was most readily accomplished by weighing the incubation vessels.

It was soon realized that a set of rigid controls was obligatory, and the following were uniformly applied:

1. The substrate was incubated in the absence of cells.
2. All of the incubations were carried out in the dark to avoid complications from photochemical transformations.
3. Periodic checks were made to ensure the viability of the cells and the absence of contaminating organisms.

The application of these measures increased both the complexity of the investigations and the analytical burden, but the disadvantages were considered to be more than compensated by the removal of possible artefacts. Additionally, virtually every experiment was repeated at least twice.

At the outset, it was decided that it was highly desirable to obtain relative rates of degradation. This meant that, for insoluble substances, the whole of the sample had to be sacrificed at each sampling time. Experiments were generally carried out in serum bottles of suitable capacity using cellulose plugs. A number of substrates, however, have appreciable vapor pressures even at room temperature, so that experiments with these required the use of a totally enclosed system. Again serum bottles were used, though in this case, they were sealed with crimp caps fitted with Teflon-covered rubber linings. It should be stressed that the Teflon must be of sufficient thickness to prevent diffusion of lipophilic substances into the rubber and that the rubber must be of sufficient thickness and elasticity to enable the caps to be tighly clamped. In extreme cases, use of sealed glass ampoules may be necessary. Since the systems are enclosed, the bottles/ampoules must be of adequate volume so that the cultures do not become totally anaerobic during incubation.

We have become increasingly aware of problems posed by the toxicity of the substrate. It was therefore decided to incorporate a test for toxicity based on the standard procedure [83] for assessing resistance to antibiotics. Cell suspensions prepared from cultures grown on a defined medium with a suitable substrate (generally 3- or 4-hydroxybenzoate) were centrifuged under sterile conditions, resuspended in buffer and spread onto plates of soy-tryptone agar. Filter paper disks impregnated with solutions of the test substrates (6–300 µg per disk) were placed on the surface and the plates incubated for up to 5 days at 30 °C. Inhibition zone diameters were measured after 2 and 5 days, because "resistance" to some substrates was induced after the longer period of incubation.

Analytical Methods

Analysis of the substrate and of metabolites is clearly a central issue. In the absence of suitably radioactive-labelled compounds, use may be made of a variety of methods, for example, ultraviolet/visible absorption, fluorescence, high performance liquid chromatography (HPLC) and gas chromatography (GC).

Since we have been extensively engaged in work with chlorinated aromatic compounds, we have used GC, most often operated in the EC mode. A number of experimental aspects seem noteworthy:

1. It is clear that essentially quantitative recovery of the substrate and/or metabolite is imperative. Solvents such as ethyl acetate, diethyl ether, di-*iso*propyl ether, methyl *iso*propyl ketone, toluene, hexane, chloroform, and dichloromethane have been used. None of these is ideal, and their capacity for dissolving substantial amounts of water must be taken into consideration. This is particularly important when GC analysis involves preparation of derivatives utilizing water-sensitive reagents (see below).

2. An unexpected difficulty arose with certain lipophilic substances such as chloroanisoles and 4-chlorobiphenyl. These appeared to be strongly adsorbed onto the glass walls of the containing vessels so that extraction by normal procedures was ineffective. Sonication after addition of acetonitrile [84] was found to overcome the problem successfully. Adsorption to the cells, particularly of Gram-positive organisms, was found to be a problem at substrate concentrations less than 1.0 mg.L^{-1}. Application of the above procedure gave a satisfactory solution.

3. Water soluble compounds (or those poorly soluble in organic solvents) presented a problem. After removal of cells by centrifugation, samples were dried. For small volumes (< 200 µL), this could be accomplished by placing them overnight *in vacuo* over P_2O_5 in a desiccator. Larger volumes (max. 300 mL) may be freeze-dried. In any of these procedures, readily volatile compounds will inevitably be lost and this should be considered.

4. This is clearly not the place to enter into details of procedures for preparation of derivatives suitable for gas chromatography, and reference should be made to several handbooks [85–87] for details. It should, however, be emphasized that no procedures for preparation of suitable derivatives are universally applicable. We have made extensive use of trichloroethyl esters (carboxylic acids), *O*-acetates, and *O*-heptafluorobutyrates (phenols) and trimethylsilyl ethers (phenols and alcohols).

Metabolic Patterns

Introduction

At this stage, an attempt is made to present a schematic outline of diverse metabolic patterns. Examples have been drawn mainly from the degradation and transformation of aromatic compounds, particularly those bearing halogen substituents.

There is an extensive literature on the biodegradation of chlorophenols, chlorobenzoates, and chloroaromatic hydrocarbons (Table 3) as well as data on the "co-metabolism" of an even wider range of chloroaromatic compounds (Table 4). The selection of examples was based not only on the specific interest of the authors, but on two other considerations:

1. The metabolism of aromatic compounds lacking substituents has been extensively studied over many years [115–121], so that details of the metabolic pathways have been delineated [122–124].

Table 3. Microbial degradation of chlorophenols, chlorobenzoates and chloroaromatic hydrocarbons

Substrates	References
Chlorophenols	[88–92]
Chlorobenzoates	[93–99]
Chlorohydrocarbons	[7, 100–103]

Table 4. Examples of co-metabolism of aromatic compounds

Chlorobenzene [104]	2,4,5-trichlorophenoxyacetate [110]
Chloronaphthalene [105]	2,3,6-trichlorobenzoate [111]
Dibenzodioxin [106]	Fluorobenzoate [112–114]
Chlorophenol [107, 108]	
4-thiomethylphenol [109]	

2. It has become generally clear that the presence of electron-withdrawing substituents (chlorine, nitro, sulphonate) hinders the degradation of the compounds (see later). Valuable general principles may then be expected to emerge
from the study of such compounds.

We have chosen data from our own investigations, because these were carried
out by uniform procedures. Nevertheless, an attempt has been made to place
these data against the background of previous studies. No attempt at an exhaus-
tive treatment has been made, and emphasis is placed upon general principles,
particularly those which appear to have the greatest environmental significance.
It is hoped, furthermore, that these principles will find wide application and that
they are not limited to compounds of restricted structural diversity.

Methods for isolating some of the strains used in the following examples have
already been described. Other strains were isolated from enrichments using
methane and methylamine [125] as sole sources of carbon. The isolates chosen for
further study were more or less randomly selected on the basis of their metabolic
versatility towards a range of aromatic, and long-chain (C_9, C_{10}) aliphatic, carboxylates. The ability to utilize methane was not taken into consideration [126].
Although most of the strains were isolated from Baltic Sea sediments, none of
them displayed an obligate requirement for NaCl. The isolates were provisionally
grouped as follows:

Gram staining	Probable taxon	Strain numbers
Positive	*Rhodococcus* sp.	1366, 1395, 1487, 1632
Negative	*Pseudomonas* sp.	1556, 1557, 1558
Negative	*Acinetobacter* sp.	1559
Negative	Unassigned	1631, 1637

The last two strains were catalase negative

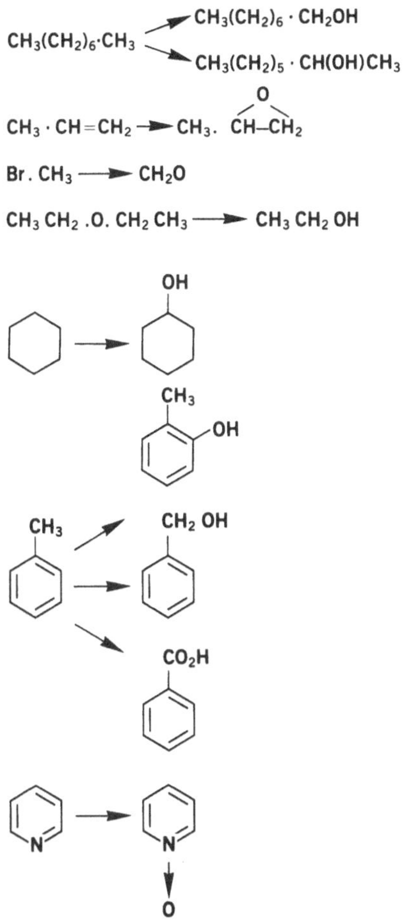

$CH_3(CH_2)_6 \cdot CH_3$ →→ $CH_3(CH_2)_6 \cdot CH_2OH$

$CH_3(CH_2)_6 \cdot CH_3$ ↘ $CH_3(CH_2)_5 \cdot CH(OH)CH_3$

$CH_3 \cdot CH=CH_2$ → $CH_3 \cdot \overset{O}{\overset{/\backslash}{CH-CH_2}}$

$Br \cdot CH_3$ → CH_2O

$CH_3 CH_2 \cdot O \cdot CH_2 CH_3$ → $CH_3 CH_2 OH$

Fig. 4. Transformations carried out by methylotrophic bacteria [130, 131, 134]

The choice of methane as the substrate for enrichment may seem, at first sight bizarre. However, an extensive literature exists [127–131] dealing with methane oxidation, and it has become clear that soluble methane monooxygenase has a re-markable substrate versatility. For example, the following activities have been demonstrated [132, 133]: hydroxylation of alkanes (C_1–C_8); oxidation of halo- and nitro-alkanes; epoxidation of alkenes; and oxidation of ethers, cyclohexane, benzene, and toluene (Fig. 4). Many of these activities have also been found in whole cells of *Methylosinus trichosporium* [134]. Especially noteworthy is the re-ported reductive dechlorination of 3-chlorotoluene to benzyl alcohol. Such wide, and relatively non-specific oxidative activity seemed ideally suited to the require-ments of the present investigation, whereas use of other substrates could well have resulted in organisms with unduly circumscribed catabolic activity. Attention is also drawn to the existence of facultatively methylotrophic bacteria [52, 135–137]. It should be noted, however, that since the capacity for growth with methane is readily lost, the obligatory requirement for methane utilization has been removed

from the most recent circumscription of the genus *Methylobacterium* [126].
Methanotrophic yeasts have also been described [138].

Exemplification

1. The simplest situation occurs when a substrate is degraded without accumulation of demonstrable intermediates. Therefore, such substances would not be considered recalcitrant.

As examples from current studies, the three isomeric monochlorophenols and 3-chloro-4-hydroxybenzoate may be given. It should be noted, that 2,4-dichlorophenol was much more resistant to degradation than any of the monochlorophenols although, with certain strains, *O*-methylated metabolites were formed (see section 5 below.

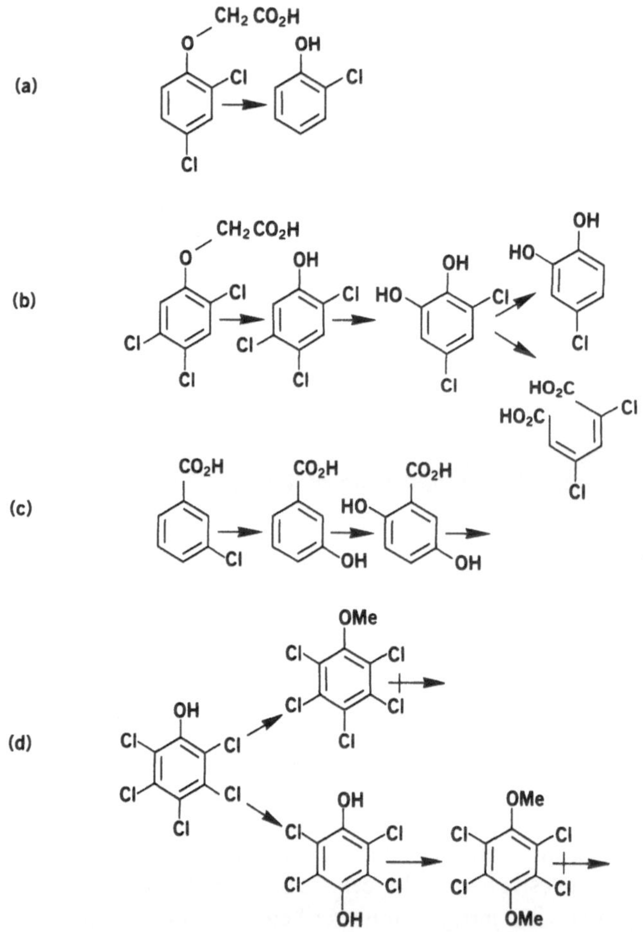

Fig. 5. (a), (b) Reductive elimination of chloride during metabolism of chlorophenoxyacetates [110, 142]. Hydrolytic elimination of chloride during metabolism of (c) 3-chlorobenzoate [143] and (d) pentachlorophenol [144]

This increased recalcitrance is consistent with the generally greater resistance to degradation of polychlorinated compounds [139, 140] and with the fact that in polycyclic monochlorinated compounds, the unchlorinated ring is attacked first [100, 101, 103, 141]. All of these observations support the operation of an electrophilic mechanism for oxidation of aromatic rings. In this context two rather unusual transformations should, nevertheless be noted: (i) the replacement of a chlorine substitutent on the aromatic ring by hydrogen in phenoxyacetates (Fig. 5 a, b); (ii) the replacement of the chlorine substitutent in 3-chlorobenzoate by a hydroxy group before further hydroxylation to 2,5-dihydroxybenzoate (Fig. 5 c), and in pentachlorophenol, with the formation of tetrachlorohydroquinone (Fig. 5 d).

2. A more complex situation arises when a metabolite is formed transiently and subsequently degraded. This may be illustrated by the case of 4-chloroguaiacol from which low concentrations of 4-chlorocatechol were transiently formed. The identity of the 4-chlorocatechol was established by its isolation and mass spectrometric comparison with the authentic compound, and its ready biodegradability confirmed in a separate experiment. Another transformation also took place which was apparently peripheral to degradation; that is, O-

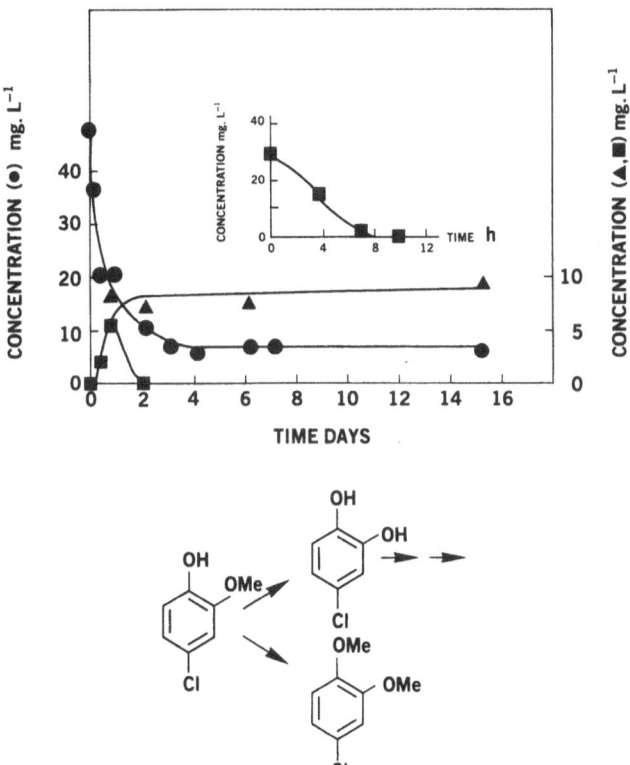

Fig. 6. Kinetics of degradation of 4-chloroguaiacol (●) and synthesis of 4-chlorocatechol (■) and 4-chloroveratrole (▲) by strain 1395. The inset shows the degradation of 4-chlorocatechol by the same strain in a separate experiment [82]

methylation also occurred (Fig. 6) which is discussed in more detail in section 5 below. Comparable observations were made during the metabolism of 4,5-dichloro, 3,4,5-, and 4,5,6-trichloro- and tetrachloro-guaiacols, although in these examples the concentrations of the corresponding catechols were low (0.3–0.6 mg.L^{-1}). Their structure was confirmed by the identity (GC) of their heptafluorobutyrates and acetates with authentic compounds. Synthesis of transient metabolites is a situation perhaps frequently encountered, though it has probably been under-documented because of inadequate analysis. However, it has been demonstrated, for example, during the degradation of 2,4-dichlorophenoxyacetate [145–146]. Because, however, a systematic study [90] showed that 2,4-dichlorophenoxyacetate was ca. 100 times less toxic than the corresponding dichlorophenol, organisms able to oxidize the phenoxyacetate would not be expected to accumulate substantial concentrations of the dichlorophenol. Therefore, the key enzymes for total degradation would be those capable of bringing about rapid metabolism of the toxic intermediate (see also section 4 below).

3. Metabolites may, of course, be more resistant to degradation, so that they persist over quite extensive periods. This may be illustrated by the results ob-

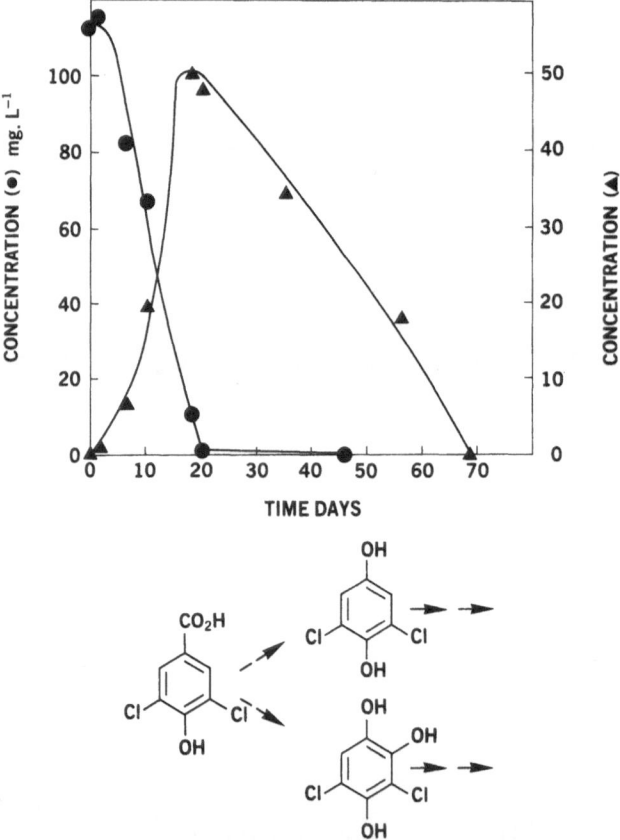

Fig. 7. Kinetics of degradation of 3,5-dichloro-4-hydroxybenzoate (●) and synthesis of a metabolite (▲) by strain 1395. The concentrations of the metabolite are given in arbitrary units

tained with 3,5-dichloro-4-hydroxybenzoate. By comparison with the corresponding monochloro compound (see 1. above), it was much more resistant to degradation. A relatively stable metabolite was synthesized which was slowly degraded during lengthy incubation (Fig. 7). This could plausibly be 2,6-dichlorohydroquinone or possibly a dichloropyrogallol. Further studies are directed to establish conclusively the structure of this metabolite.

4. Numerous studies (Table 3,4) have examined the metabolism and cometabolism of halogenated benzoates, whereby an unexpectedly complex situation has emerged. Results from the recent studies have been selected to illustrate some particularly interesting facets of this type of metabolism. In order to simplify the discussion, the expected metabolic patterns and the resulting products have been summarized in Fig. 8. The actual situation was quite complex. Incubation of cell suspensions of the three isomeric monochlorobenzoates resulted in their relatively rapid disappearance (20–40 h). Metabolites were synthesized simultaneously, and these were persistent for up to at least 30 days. Although they would have been expected to be 3- and 4-chlorocatechols, they were not identical with samples of authentic reference compounds. Possibly they are monochlorohydroquinones or monochloropyrogallols, and further quantities of material have been prepared for determination of the structure. Analogous formation of nitrohydroquinone has been demonstrated during metabolism of 3-nitrophenol by a flavobacterium [147]. By analogy with the reported [148] inhibition of catechol 2,3-dioxygenase activity by 3-chlorocatechol, it was speculated that a similar situation might hold

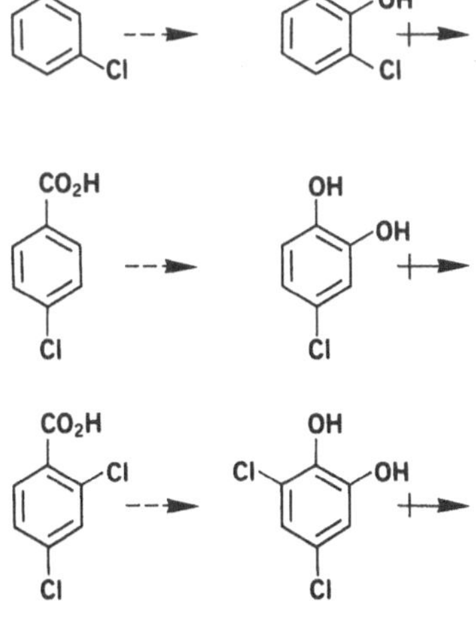

Fig. 8. Schematic outline of degradation of 2-,3-,4-chloro- and 2,4-dichlorobenzoates

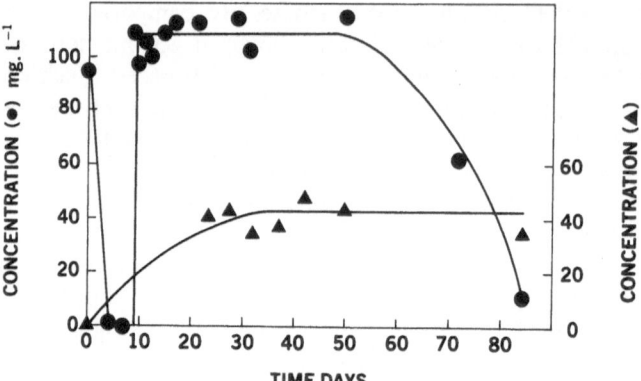

Fig. 9. Kinetics of degradation of 2,4-dichlorobenzoate (●) and synthesis of a metabolite (▲) by cells of strain 1366 which had already totally degraded the substrate. Metabolite concentrations are given in arbitrary units. Cells were incubated with the substrate which was totally degraded in 6 d.; fresh substrate was added to give a concentration equal to that used initially

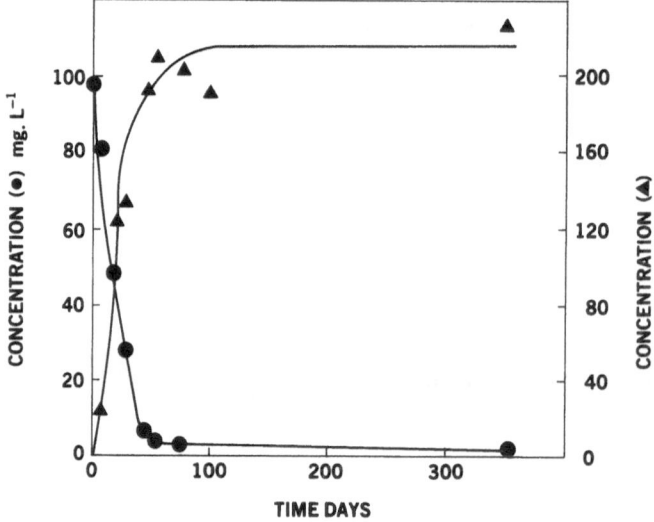

Fig. 10. Kinetics of degradation of 2,4-dichlorobenzoate (●) and synthesis of a metabolite (▲) by strain 1395. Concentrations of the metabolite are in arbitrary units

for these metabolites. Consequently, degradation of the chlorobenzoates would be inhibited by their metabolites. This expectation was indeed fulfilled as shown by the results from the experiment where fresh substrate was added to a culture which had already totally degraded 2-chlorobenzoate. There was a significant lag phase before degradation of the fresh substrate, and this then occurred at a much lower rate than that found initially. This was demonstrated even more dramatically with 2,4-dichlorobenzoate, and the results are shown in Fig. 9.

An additional feature was the low residual concentration of the initial substrate even after prolonged incubation (Fig. 10).

It may be further noted, that cells capable of the total degradation of halogenated benzoates have evolved a number of ways to avoid the synthesis of toxic metabolites. For example, in strains able to utilize 3- and 4-chlorobenzoates, this is accomplished by the synthesis of catechol 1,2-dioxygenase instead of the 2,3-dioxygenase which would result in the subsequent synthesis of toxic chlorocatechols [9]. This avoidance of toxic-metabolite synthesis is discussed in more detail later.

Collectively, these results support the view, that metabolites may be synthesized which are not only resistant to further degradation by the organism producing them, but which may even inhibit degradation of the parent substrate. Although in certain laboratory strains the selective pressure has resulted in organisms able to circumvent such undesired metabolic consequences, the frequency of their occurrence in natural situations is totally unknown.

5. In all of the previous examples, metabolism resulted in the formation of various hydroxylated metabolites, some of which were degraded further. This situation appears to be the one most commonly encountered. The synthesis of lipophilic metabolites which were seemingly resistant to further transformation was, therefore, unexpected. This was encountered, however, during the metabolism of polychlorinated phenols and guaiacols. For all of the compounds which were examined, O-methylation to anisoles and veratroles was observed. In view of the significance of these observations, the unambiguous confirmation of the structure of the metabolites was deemed necessary. In all cases, this was carried out by mass-spectrometric comparison of the metabolites with authentic reference compounds. The yields of these metabolites depended both on the substrate used (Table 5) and on the strain (Table 6). Current investigations have shown that the metabolite yields are also critically dependent upon the concentration of the substrates.

Significantly different results were obtained in experiments with the two isomers of trichloroguaiacol examined. Transformation of the 4,5,6-isomer produced a high yield of the corresponding veratrole (Fig. 11). For the 3,4,5-isomer,

Table 5. Concentrations (mg · L^{-1}) of metabolites produced by strains 1395 and 1487 from corresponding guaiacols and phenols (50 mg · L^{-1}) [82]

Metabolite	Strain 1395	Strain 1487
4-chloroveratrole	9.0	28.0
4,5-dichloroveratrole	0.5	0.6
3,4,5-trichloroveratrole [a]	0.5	0.15
3,4,5-trichloroveratrole [b]	13.0	11.0
Tetrachloroveratrole	0.9	1.0
3,4,5-trichloro-2,6-dimethoxyophenol [a]	7.0	ND
2,4,6-trichloroanisole	9.6	11.0
Pentachloroanisole	1.0	0.15

[a] From 3,4,5-trichloroguaiacol
[b] From 4,5,6-trichloroguaiacol
ND Not detected

Table 6. Yield (%) of 3,4,5-trichloroveratrole produced by various bacteria from low (100 µg · L^{-1}) and high (50 mg · L^{-1}) concentrations of 4,5,6-trichloroguaiacol. Strains 1395 and 1632 are gram-positive organisms; all the others are gram-negative

Strain	Low conc'n	High conc'n	Ratio (low/high)
1395	100	26	3.8
1632	100	76	1.3
1556	9	0.3	30
1557	30	0.7	43
1558	21	0.4	52
1559	56	1.8	31
1631	47	1.0	47
1637	71	1.3	55

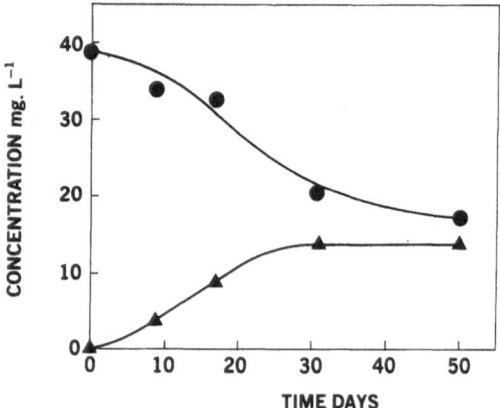

Fig. 11. Kinetics of degradation of 4,5,6-trichloroguaiacol (●) and synthesis of 3,4,5-trichloroveratrole (▲) by strain 1395

however, a considerably more complex picture emerged (Fig. 12), whereby three principal metabolites were produced:
1. in lower yield, 3,4,5-trichloroveratrole
2. in substantially higher yield, 3,4,5-trichloro-2,6-dimethoxyphenol produced presumably by hydroxylation and subsequent *O*-methylation at C 6
3. 3,4,5-trichlorocatechol which had only a transient existence and was rapidly degraded.

Although under the conditions of these experiments the 3,4,5-trichloro-2,6-dimethoxyphenol was resistent to further transformation, it has been shown that, at lower substrate concentrations, it was virtually quantitatively transformed into 1,2,3-trichloro-4,5,6-trimethoxybenzene. At high concentrations, the transformation was greatly facilitated by the presence of a co-substrate [81]. A reaction analogous to 2. above has been demonstrated: 7-ethoxy-coumarin was transformed by a cell suspension of *Streptomyces griseus* into a mixture of 6-hydroxy-7-methoxy and 6-methoxy-7-hydroxy-coumarin (Fig. 13).

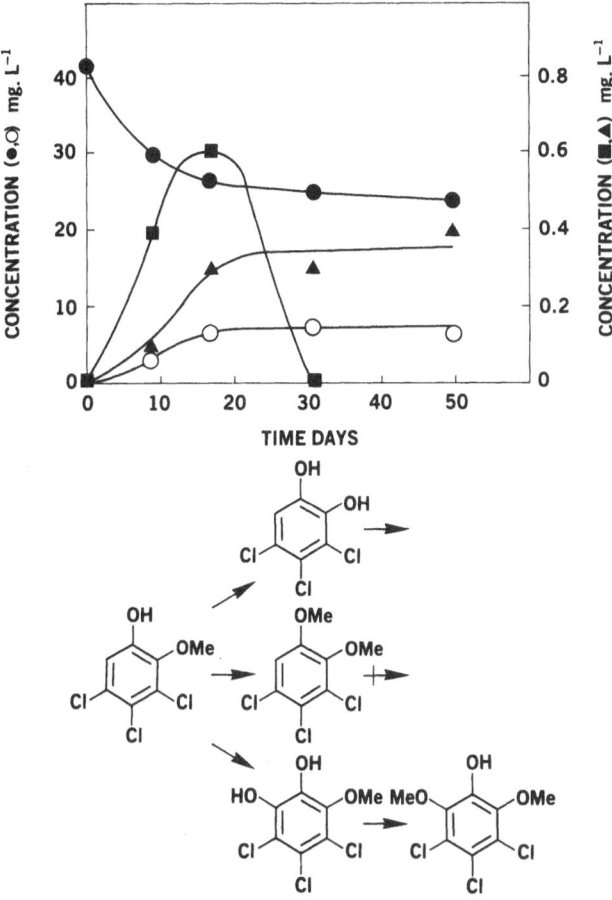

Fig. 12. Kinetics of degradation of 3,4,5-trichloroguaiacol (●) and synthesis of 3,4,5-trichloroveratrole (▲), 3,4,5-trichlorocatechol (■) and 3,4,5-trichloro-2,6-dimethoxyphenol (○) by strain 1395 [82]

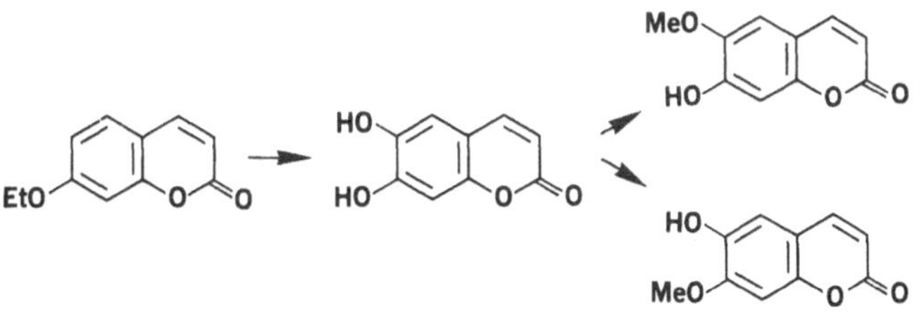

Fig. 13. Transformation of 7-ethoxycoumarin with formation of 6-methoxy-7-hydroxy- and 6-hydroxy-7-methoxycoumarin by *Streptomyces griseus* [149]

Bacterial O-methylation of pentachlorophenol has already been described [144, 150], though the yields were extremely low (0.005 to 0.02%). Moreover, O-methylation of both pentachlorophenol [151] and of tetrachlorophenol [152] has been established in a number of fungi. In general, however, microbial O-methylation seems to be more restricted than it is in mammalian systems [153], although catechol O-methyl transferase has been demonstrated in some species of *Candida* [154] and the enzyme has been purified [155]. It is, however, clearly premature to speculate further on the enzymatic basis of the present findings. Attention should also be drawn to the synthesis of *iso*butyraldehyde oxime O-methyl ether from valine [156] which is presumed to involve O-methylation of an intermediate hydroxylamine by an enzyme distinct from catechol O-methyltransferase.

There did not appear to be any correlation between the relative contributions of O-methylation and degradation so that these methylations may be regarded as tangential to the normal metabolism of the cell. Presumably, they function as a detoxification mechanism to the cell producing them. This is supported by the data on toxicity of the chloroguaiacols and pentachlorophenol: all were toxic though to varying degrees. In order of increasing toxicity: (3,4,5-trichloro-2,6-dimethoxyphenol = 4,5-dichloroguaiacol = 4,5,6-trichloroguaiacol) <3,4,5-trichloroguaiacol < pentachlorophenol. By contrast, the veratroles displayed no toxicity with the exception of 4,5-dichloroveratrole which was, however, only weakly toxic in comparison with the guaiacols. Such results provide strong support for the view that O-methylation is a detoxification mechanism, particularly since a number of Gram-negative organisms which were virtually resistant to the above compounds displayed a very much weaker methylation capacity (Table 6). Additional support for this view is provided from the fact that while experiments using substrate concentrations of 50 mg.L^{-1} showed product yields from 1% to 50%, similar experiments at lower concentrations (100 µg.L^{-1}) showed virtually quantitative conversion within periods varying from 6 h to 20 h. The greatest effect of concentration was observed with Gram-negative bacteria which were much less sensitive to the substrates examined.

It is interesting to observe that Gram-negative bacteria are generally resistant to erythromycin and contain methylated adenine in their 23 S rRNA [157]. Resistance to erythromycin in Gram-positive bacteria results in N^6, N^6 dimethylation of adenine in the 23 S ribosome [158]. Whereas all of the Gram-positive strains examined in this study were sensitive to erythromycin, the Gram negative strains were uniformly resistant.

Determinative Factors

Biodegradability is a function of many factors. It is therefore appropriate to discuss some of these in more detail by illustrating the possible vulnerability of laboratory tests of biodegradation.

Physical Aspects

Temperature, Salinity, pH, Oxygen Tension. Two situations may be distinguished:
1. These parameters are stringently applied, so that only obligately adapted organisms are significant (psychrophiles, thermophiles, marine bacteria, freshwater bacteria, acidophiles, alkalophiles, anaerobes).
2. These factors affect only in degree the activity of the test organisms. For example, mesophiles will generally grow and metabolize slowly even at low and

high temperatures; many organisms though inhibited by high concentrations of NaCl (>1 M) grow readily at lower concentrations; and strictly aerobic organisms may grow only poorly under conditions of low oxygen tension.

Clearly then, all of these parameters must be taken into consideration in assessing the environmental significance of laboratory results. It has become obvious, however, that unexpected complications may arise. For instance, many anaerobes are relatively tolerant of oxygen. Moreover, what is more surprising is, that strictly aerobic organisms, maintained under conditions of low oxygen tension, synthesize enzymes typical of an anaerobic way of life [159]. Such observations are especially relevant to metabolic transformations occurring in sediments with low oxygen tensions. Transformations carried out by facultatively and strictly anaerobic bacteria, and by aerobic bacteria under anaerobic conditions using a variety of electron acceptors such as nitrate, are discussed later.

The Concentration of the Substrate. The concentrations of many compounds after discharge into the environment may be extremely low. The importance of rates of degradation at low concentrations must therefore be evaluated. Four aspects seem especially noteworthy:

1. Investigations of biodegradation by natural microbial communities have shown two distinct patterns:
   - rates of degradation were a linear function of concentration over a wide range (pg.mL^{-1} to ng.mL^{-1}) [160, 161].
   - rates of degradation at low concentration were extremely low so that a threshold concentration appeared to exist whereby such compounds would become "persistent" [162].
2. Extensive investigations of the bacterial flora of natural waters have revealed the presence and significance of oligotrophic bacteria able to utilize extremely low concentrations of substrate [37, 38]. Although such organisms will not normally be isolated by conventional isolation procedures, they may be highly significant in natural communities.
3. The existence of bacteria in laboratory distilled and deionized water is well known. It has now been shown that commonly encountered bacteria such as *Aeromonas hydrophila* [163] and *Pseudomonas aeruginosa* [164] may effectively scavenge extremely low concentrations of substrate and may therefore be found in ostensibly "substrate-free water." For the pseudomonad, there was apparently no correlation between the capacity to use the substrates at low concentrations and at concentrations 1000-fold higher. It has also been shown that with a flavobacterium [165], the substrate versatility at low concentrations was appreciably more restrictive. Furthermore, for motile organisms in natural situations, their chemotactic response may play an important role.
4. At low substrate concentrations, there may be an alteration in the regulation of catabolic enzymes [166]. This is briefly noted later.

Chemical Aspects

Bioavailability

In many natural waters and particularly in sediments, substrates may be bound to a variety of macromolecules such as polysaccharides, proteins, nucleic acids, lipids, and humic material. Although the degree of binding varies widely, the covalent binding of, for example, pesticides to humic material in the terrestrial system has been clearly established [167, 168]. This results in the apparent reduction in the "available concentration" and a concomitant increase in concentrations which may be only slowly mobilized and released. Substantial activity has therefore been devoted to such problems which are significant enough to justify more than a passing reference here.

In the terrestrial system which has been most thoroughly investigated, it appears that a combination of chemical, biological, and enzymatic processes is important. An example of the last is the role of fungal phenol oxidases in cross-linking substrates to structural elements in humic material [169].

The structure of humus is extremely complex and is unknown in detail. It is a heterogeneous substance, and terrestrial and aquatic humus material differ in structure [170]. At least the major structural entities of humus have been established, so that a number of plausible mechanisms for their reaction with xenobiotics or their metabolites may be suggested [171, 172]. These include nucleophilic addition, electrophilic substitution, reaction with active methylene groups and Diels-Alder reactions (Fig. 14), and all of them involve interactions between reactive functional groups. It has, however, been shown that purely adsorption processes may also be important with neutral molecules such as hexachlorobenzene [173] and polycyclic aromatic hydrocarbons [174]. The significance of such interactions may be illustrated from the reduced rates of biodegradation of the octyl ester of 2,4-dichlorophenoxyacetate in the presence of humus [84].

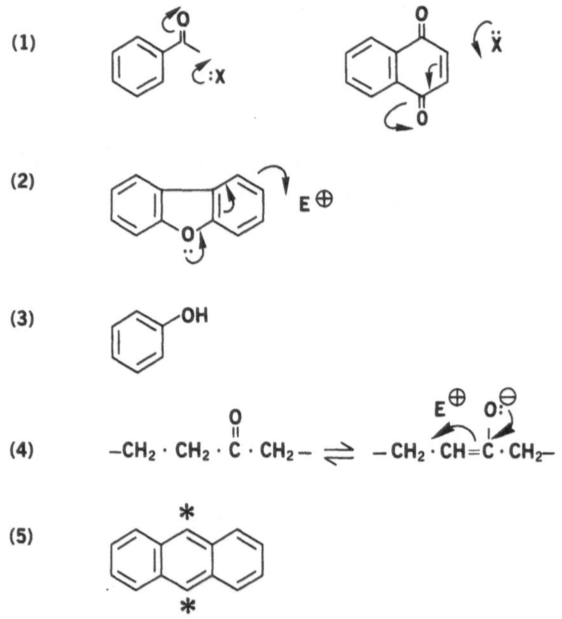

Fig. 14. Schematic outline of possible reactions involving structural elements in humus. The asterisks show the positions at which Diels-Adler condensation might occur

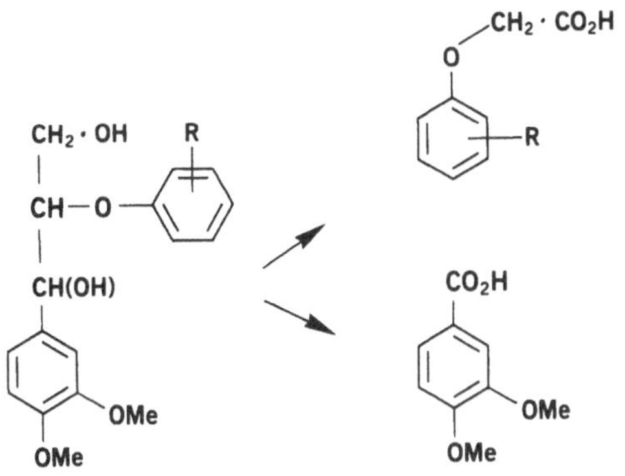

Fig. 15. Metabolism of substituted phenols (R = H) chemically bound to a model lignin structure [175]

A limited number of studies have been directed to studying the microbial "mobilization" of substances bound to simplified models of the structure of lignin (Fig. 15).

Completely different mechanisms operate for interaction between organic compounds and various types of minerals [176], but again such binding results in a diminution in the bio-availability of the original compound.

All of these processes result in a reduction in the concentration of the substance available for microbial attack. While this represents a measure of detoxification towards the microorganisms, it means that estimates of biodegradability which do not adequately take these processes into consideration, may grossly underestimate the persistence of the compounds.

Mechanisms also exist whereby the microorganisms themselves *increase* the bioavailability of the substrate. For example, microorganisms growing at the expense of insoluble hydrocarbons have been shown to synthesize emulsifying agents [177, 178] which apparently result in increased accessibility of the substrate to the organisms. These emulsifying agents have varying structures but generally consist of a carbohydrate component (trehalose [179, 180], rhamnose [181] or a heteropolysaccharide [182]) esterified with long-chain fatty acids which may carry methyl or hydroxyl side chains.

The plausible role of these substances, however, is challenged by the finding that *Torulopsis petrophilum* may produce either a surfactant or an emulsifier, and that these were not produced to facilitate uptake of insoluble substrates [183].

The Structure of the Substrate

The structure of the substrate is of major importance in determining its biodegradability. Strenuous efforts have therefore been made to produce empirical rules which are of predictive value in assessing relative biodegradability. Two possible approaches are presented.

Structure-activity relationships. Brief mention must be made of these, as they have been already successfully applied to abiotic processes. However, it is not clear that structure-activity relationships may be equally and effectively applied to biological reactions such as those involved in biodegradation. There are significant differences between the two areas of application: biodegradation (as opposed to biotransformation) involves a series of consecutive reactions of which only the first is determined by the structure of the parent substrate. Since the Hansch approach [184] has been valuable particularly in situations where a series of structurally related substances was to be evaluated (e.g. plant hormones, pharmaceuticals), it may be useful for restricted problems in biotransformation. A good example is provided by a study of the transformation of various substituted phenols into the corresponding catechols [185].

Mechanistic considerations. This is not the appropriate place for detailed discussions of the mechanisms of biodegradation. Nevertheless, a mechanistic viewpoint is justified for two reasons:
1. It provides a simple basis for categorizing a wide range of biotransformations.
2. It gives a rational explanation for the synthesis of possibly unexpected metabolites on the basis of plausible pathways for their production.

The majority of biotransformations carried out under aerobic conditions involve oxidations catalysed by a gamut of oxidative enzymes. Their activity will be illustrated for two large groups of compounds with greater emphasis placed on those compounds generally considered recalcitrant, namely; 1. alkanes and cycloalkanes and 2. aromatic compounds.

1. Alkanes and cycloalkanes. n-alkanes are metabolized by oxidation at the α-, β-, or ω-positions, followed by dehydrogenation to aldehydes (or ketones) and oxidation to carboxylic acids. These are then degraded sequentially by β-oxidation with cleavage of acetate units. The possible physiological role of ω-oxidation has been discussed [186]. A number of mechanisms are available for degradation of propionate produced from odd-numbered alkanes [187]. Two structural factors result in less readily degraded compounds:
– substitution by halogens [188, 189] and
– branching of the aliphatic chain [190]

Certain positions of branching are clearly most critical in hindering β-oxidation, though carboxylation mechanisms may be evolved in order to relieve this blockage [191]. Although degradation of cycloalkanes by pure cultures has been demonstrated [192, 193], it appears that such compounds are generally recalcitrant. It may also be noted that hydroxylation of n-alkanes by a cytochrome P-450 system has been demonstrated in yeasts [194], and occasionally in bacterial systems degrading camphor [195] and octane [196]. For the sake of completeness, it may be noted that, while metabolism of complex alkenes proceeds by cleavage of the double bond with formation of aldehydes or ketones [197], metabolism of simple acetylenes and acetylene carboxylates involves initial hydration followed by isomerization of the vinyl alcohol [198].

2. Aromatic compounds. It has generally been established that, in prokaryotic cells, oxidation proceeds via a dioxygenase which results in the initial formation of a *cis* 1,2-dihydro-1,2-diol [199–203]. In eukaryotic organisms and in mammalian systems, however, a mono-oxygenase coupled to the cytochrome P-450 complex is generally implicated. For aromatic compounds this results in the initial

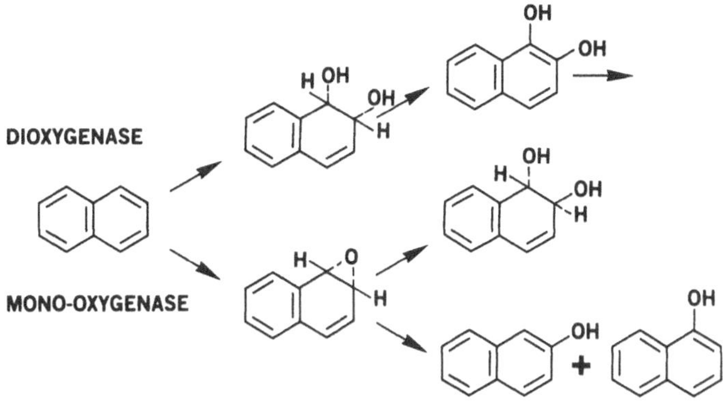

Fig. 16. Metabolism of naphthalene via a dioxygenase and a monooxygenase system showing the different products formed

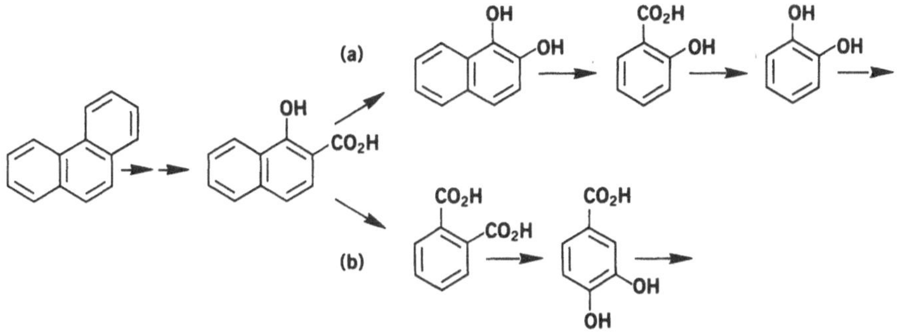

Fig. 17 a, b. Pathways adopted during the bacterial metabolism of phenanthrene : (a) [118] : (b) [120, 124]

formation of an epoxide which may subsequently be hydrated to a *trans* 1,2-di-hydro-1,2-diol [204, 205] (Fig. 16).

Relatively few studies have been devoted to the microbial degradation of poly-cyclic hydrocarbons. On the basis of these studies [118, 120, 124], however, it appears that, beyond naphthalene, linear as opposed to angular annelation decreases the ease of degradation. Moreover, divergent pathways have been demonstrated for the degradation of phenanthrene (Fig. 17).

It is not merely a matter of academic interest whether oxidation occurs via a mono-oxygenase or a di-oxygenase since there are two important consequences resulting from the oxidation route which is followed:

1. If an intermediate epoxide is formed, this may react with anions in the medium. This reaction has been suggested [56] as the source of nitrohydroxylated products formed during the metabolism of 4-chlorobiphenyl in the presence of nitrate (Fig. 2).
2. The possibility of intramolecular rearrangements (NIH shift) is opened up [206] and may be represented schematically in Fig. 18. This results in the pro-

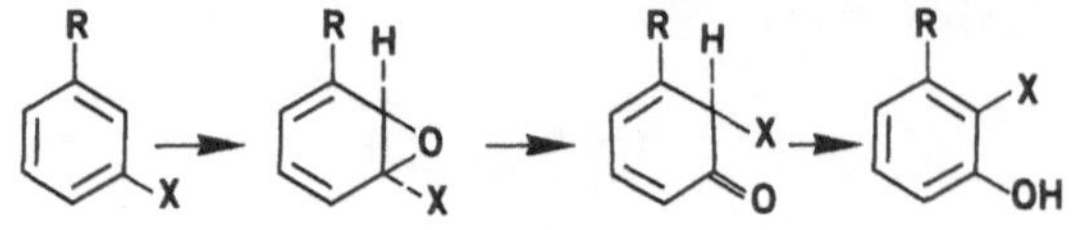

Fig. 18. Schematic outline of the NIH shift

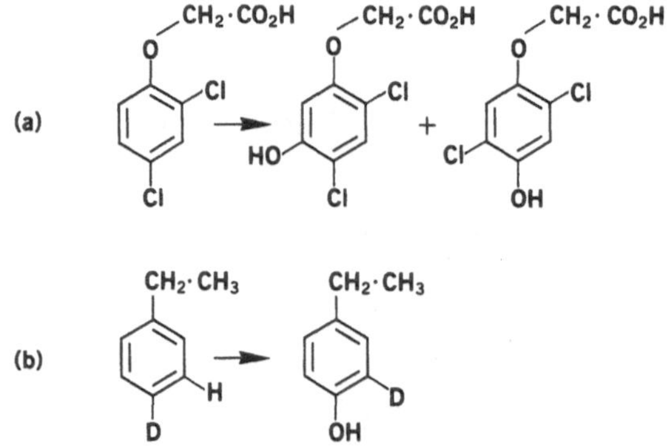

Fig. 19 a, b. Operation of the NIH shift during (a) the fungal metabolism of 2,4-dichloro-phenoxyacetate by *Aspergillus niger* [207] and (b) the metabolism of 4-^{2}H-ethylbenzene by the oxygenase system from *Methylococcus capsulatus* [209]

duction of a rearranged, as well as the expected product. This mechanism was elucidated using deuterium-substituted substrates (X=D), and indeed use of this isotope is generally obligatory. The rearrangement has, however, also been shown to occur with chlorine and alkyl groups (X=Cl, alkyl). Examples of this rearrangement have been found both in fungal [204, 207] (Fig. 19a) and bacterial [208, 209] (Fig. 19b) systems. However, the frequency with which re-arrangement occurs is unknown but would be expected during reactions in-volving attack of aromatic rings by mono-oxygenases.

In fungi and yeasts, further degradation of the *trans*-dihydrodiol or phenols produced from complex aromatic hydrocarbons does not generally occur (see later). In bacteria, however, the catechol undergoes further reactions, whereby ring cleavage may occur utilizing a 1,2-oxygenase (*ortho* cleavage) or a 2,3-oxy-genase (*meta* cleavage). The products of these reactions clearly differ, and which alternative is adopted depends both on the organisms involved and on their regu-latory mechanisms. This is briefly discussed later. Its particular significance in the degradation of halogenated aromatic compounds is taken up in greater detail be-low.

Substitution in aromatic rings by electron-withdrawing substituents such as halogen, nitro, and sulphonate substituents [210–212] results in a decrease in the ease of degradation. This may plausibly be rationalized on the assumption that dioxygenation is an electrophilic reaction.

The effect of chlorine substitution on biodegradability is nicely shown by data from experiments on a variety of polychlorinated biphenyls [213]. The relative

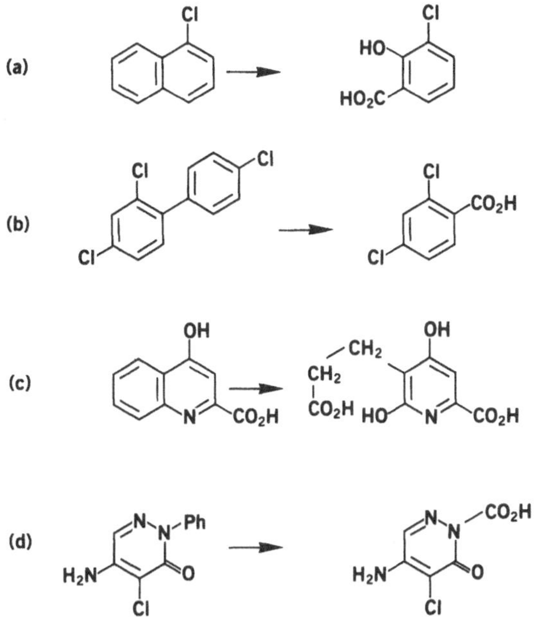

Fig. 20 a–d. Metabolism of chlorinated aromatic and heterocyclic compounds illustrating preferential oxidation of the less heavily substituted ring. (a) [100] : (b) [140] : (c) [214] : (d) [215]

ease with which substituted and unsubstituted rings are degraded is illustrated by the examples given in Fig. 20.

The metabolic pathways followed during the biodegradation of chloro, nitro, and sulphonato aromatic compounds differ significantly and will be briefly summarized in the following paragraphs.

The biodegradation of halobenzoates clearly illustrates the importance of the pathway chosen for their metabolism. The choice of the *ortho* [1, 2] or *meta* [2, 3] cleavage pathway is particularly critical for the following reason: synthesis of the 2,3-dioxygenase would result in the production of toxic and inhibitory metabolites. This may, however, be effectively avoided by implicating the 1,2-dioxygenase (Fig. 21). Complete degradation of the chlorobenzoates can therefore only occur in strains with a functional catechol 1,2-dioxygenase. Such results underline the possibly hitherto underestimated significance of the choice between the *ortho* (catechol 1,2-dioxygenase) and the *meta* (catechol 2,3-dioxygenase) pathways.

In all of these reactions, the chlorine atom is retained on the ring until cleavage has occurred. Then it is eliminated from the appropriate chloromuconate as chloride. A number of other strategies have, however, been adopted. The co-metabolism of 2-fluorobenzoate has been described and resulted in the formation of both catechol and 2-fluorocatechol [113]. Although this was recognized many years previously [216], its significance was realized only much later [217]. Strains utilizing the fluorobenzoate lack dihydrobenzene dehydrogenase activity. Thereby, by simultaneous elimination of fluoride and CO_2, the cells escape the consequences of the lethal metabolites which would have been synthesized if 1,2-

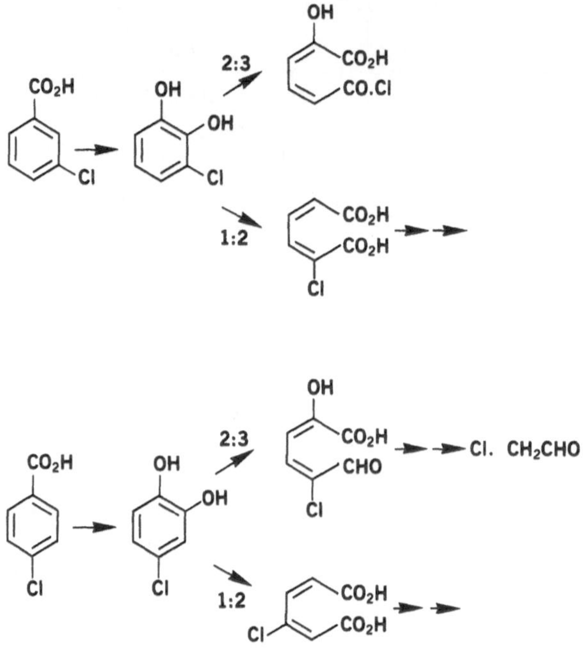

Fig. 21. Metabolism of 3- and 4-chlorobenzoates showing the products resulting from the action of catechol-1,2- and catechol-2,3-dioxygenase [9]

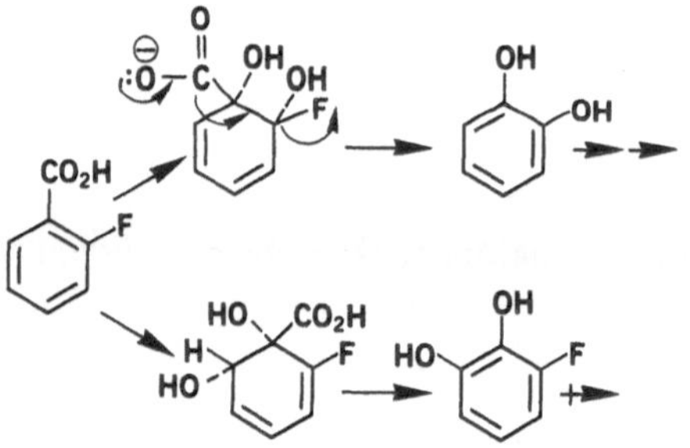

Fig. 22. Metabolism of 2-fluorobenzoate [113, 216, 217]

dioxygenation had taken place (Fig. 22). Not all cells display such metabolic intelligence, and toxic or inhibitory metabolites are not infrequently produced. Chlorine may occasionally be replaced directly by a hydroxy group (Fig. 5c) or eliminated by a reductive process (Fig. 5a, b).

The sulphonate group is, by contrast, eliminated at an early stage [61, 218, 219], so that the substrate for ring cleavage is free of inhibitory substituents (Fig. 23).

Although less detailed mechanistic studies have been carried out with aromatic nitro compounds, their metabolism has revealed a complex array of interesting metabolic possibilities. Some examples are given in Fig. 24. As would be

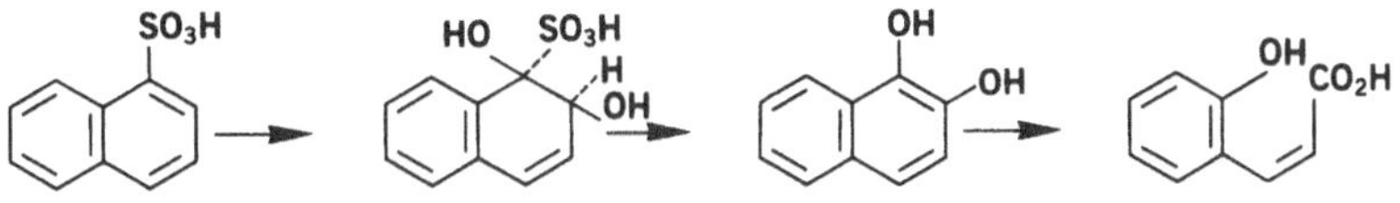

Fig. 23. Pathway for the metabolism of naphthalene-1-sulphonate [219]

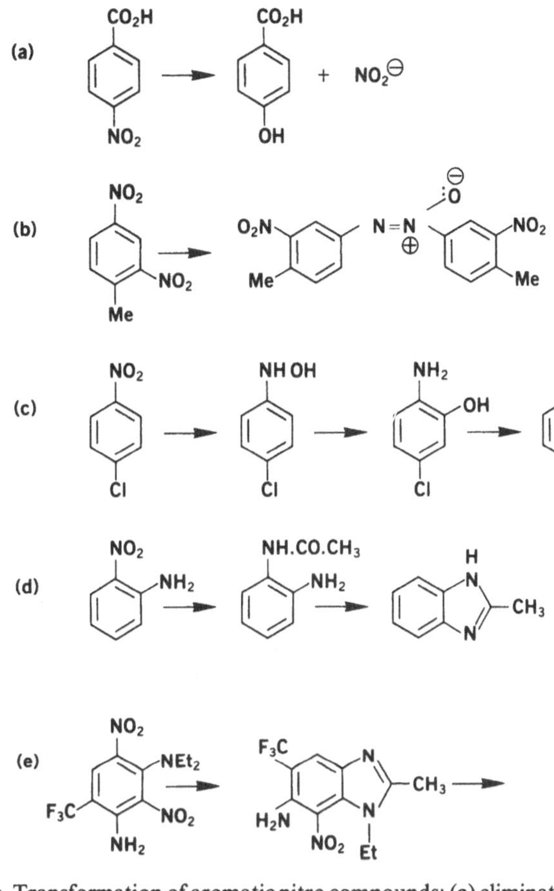

Fig. 24 a–e. Transformation of aromatic nitro compounds: (a) elimination of nitrite [220], (b) formation of azoxy compounds [221], (c) synthesis of 2-hydroxyanilines and subsequent acetylation [222], (d) cyclization of intermediate 2-aminoacetanilide with formation of benziminazole [223], (e) formation of benziminazole via oxidation of an N-ethyl group and cyclization with a vicinal amino group [224]

expected, the transformations corresponding to those in Figs. 21 b and 21 c have also been observed in chloroanilines [225].

It should also be observed that the presence of the electron-withdrawing carboxyl group does not appear to result in an increased degree of recalcitrance comparable to that introduced by halogen, nitro or sulphonate groups. Possibly a concomitant decarboxylation increases the electron density on the ring, so that electrophilic dioxygenation is facilitated. This is supported by the fact that most

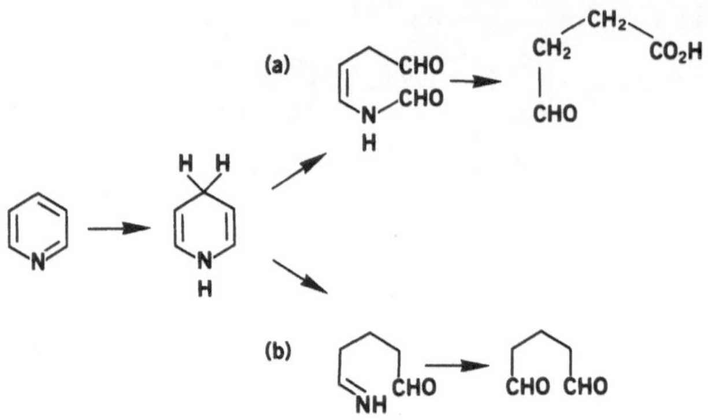

Fig. 25. Reductive pathway for the metabolism of pyridine [233]

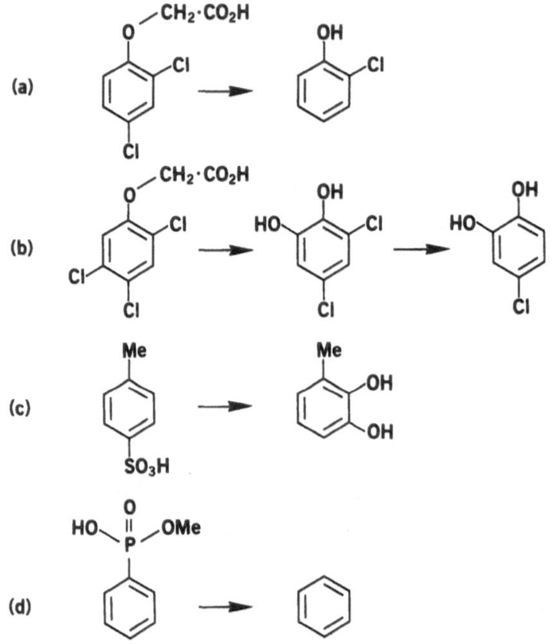

Fig. 26. Reductive elimination of substituents : (a), (b) chloride [110, 142], (c) sulphite [61], (d) methylphosphite [234]

simple aromatic (benzene, naphthalene) and heterocyclic (pyrrole, furan, pyridine [226]) carboxylates are relatively readily degraded. Indeed decarboxylation may precede ring cleavage with [227, 228] or without [229] hydroxylation.

Some additional mechanistic points may be briefly noted:

1. In bacteria able to metabolize 1,4-dihydroxybenzenes, the *meta* pathway may diverge giving products of different stereochemistry. These pathways differ in their dependence on glutathione [230].
2. Hydroxylation has occasionally been demonstrated in the degradation of heterocyclic carboxylates, e.g. pyridine [231] and furan [232]. It should be ob-

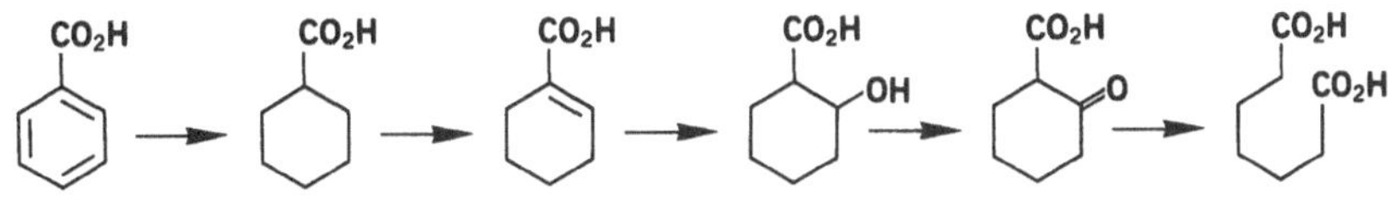

Fig. 27. Reductive pathway for the metabolism of benzoate [235, 236, 237]

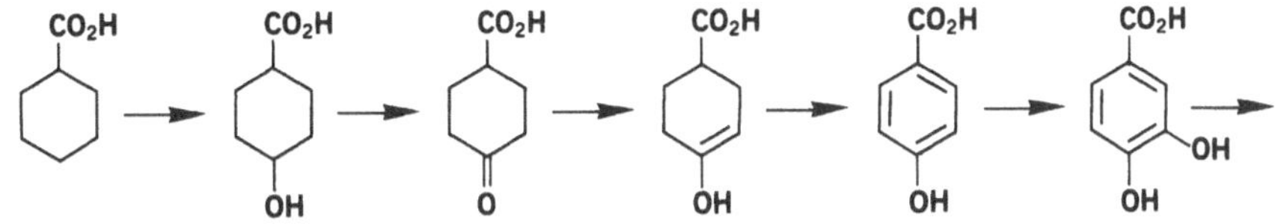

Fig. 28. Metabolism of cyclohexanecarboxylate involving dehydrogenation to 4-hydroxybenzoate [242–244]

served that an unusual aerobic degradation of unsubstituted pyridine by partial reduction of the ring has also been observed (Fig. 25).

3. Occasionally, electron-withdrawing substituents such as chlorine, sulphonate, and phosphonate may be replaced by hydrogen (Fig. 26) in formally reductive reactions whose mechanisms have not hitherto been resolved.

4. Anaerobic degradation of aromatic rings proceeds understandably by a set of reductive reactions (Fig. 27) both in the absence [235–237] and presence [238–241] of nitrate. What is more surprising, is that the *aerobic* metabolism of cyclohexane carboxylates may involve initial dehydrogenation to benzoates followed by the expected route for their further catabolism (Fig. 28).

Microbiological Aspects

Biodegradation Agents

This is neither a list of organisms known to be significant in biodegradation, nor an attempt to evaluate those of the greatest potential significance. Rather, it is an attempt to draw attention to relatively large groups of organisms whose role in the degradation of organic compounds in the environment may have been underestimated. Although some groups, such as extreme thermophiles or barophiles have quite restricted relevance, their existence should not be forgotten. Organisms belonging to taxa with well established nutritional versatility, have relatively seldom been employed in studies of biodegradation. An exception is *Pseudomonas cepacia* which has been shown to effectively degrade 2,4,5-trichlorophenoxyacetate [245].

In this review, the greatest attention has been devoted to bacteria and significantly less attention to yeasts and fungi. The eukaryotic organisms display considerable metabolic versatility and, in terrestrial systems for example, they may be the most significant organisms. Our primary interest lies, however, in aquatic systems in which bacteria are both numerically and metabolically the organisms of greatest importance.

It should also be recognized that although yeasts and fungi are capable of metabolizing a variety of aromatic hydrocarbons, they are unable to use them as sources of carbon and energy. Fungal metabolism of naphthalene [205, 246], anthracene [247], biphenyl [248, 249], and methylbenz-[a]-anthracenes [250, 251] has been studied, and it has been shown that yeasts are able to transform naphthalene, biphenyl, and benzpyrene [252]. All of these transformations should, however, be viewed as the operation of a detoxification system for the cells. This interpretation is supported, for example, by the isolation of sulphate esters and glucuronides of 1-naphthol and of 4-hydroxybiphenyl produced from the metabolism of naphthalene and biphenyl by *Cunninghamella elegans* [253]. The sulphate esters of 4-hydroxybiphenyl and of 4,4'-dihydroxybiphenyl have been shown to be produced from biphenyl by *Aspergillus toxicarius* [254]. These observations, combined with the demonstration that in fungi, aromatic ring oxygenation proceeds via a cytochrome P-450 mediated monooxygenation [246, 255], support the view that fungal systems provide a suitable model for mammalian detoxification systems [204, 256]. In the above examples, ring cleavage does not apparently occur. However, it should be noted that cleavage of simpler aromatic rings (phenols, benzoates) by both yeasts [257, 258] and fungi [259] has been demonstrated.

1. In spite of the intense interest in the microbial degradation of oil in the marine environment [23], surprisingly little work seems to have been devoted to studies in degradation using marine bacteria. This seems regrettable as it makes impossible the evaluation of the persistence of compounds which ultimately reach the open sea. It should be observed, however, that taxonomic studies with marine bacteria [260] which have relied upon their ability to utilize an extensive range of compounds as sole source of carbon and energy, suggest that marine bacteria do, in fact possess metabolic versatilities comparable to these of many terrestrial and freshwater groups such as the pseudomonads and actinomycetes.

2. Strictly aerobic bacteria may, under anaerobic conditions, make use of a number of electron acceptors other than oxygen. Of these, the most extensively studied is nitrate [54], though it has been demonstrated that trimethylamine-N-oxide may function in a comparable manner [261, 263]. This has also been demonstrated with certain oxidized sulphur compounds. For example, tetrathionate is able to support the growth of facultatively anaerobic bacteria on non-fermentable substrates [264, 265]. Moreover, dimethyl sulfoxide may play a similar role [266, 267]. In areas which contain any of these electron acceptors, their role in promoting anaerobic degradation should not therefore be underestimated. Furthermore, interesting examples have been given of degradations by denitrifying bacteria under anaerobic conditions. These include reductive transformation of aromatic rings [238–240], decarboxylation of benzoates [268] and demethylation of methoxybenzoates [269]. In certain cases, however, e.g. di-n-butylphthalate, no degradation of the ring could be demonstrated [241]. Although fumarate may also serve as a suitable electron acceptor under anaerobic conditions [270], this is perhaps of lesser environmental significance.

3. Among the common bacteria, the facultative anaerobes seem to have been largely neglected. This is surprising in view of the demonstration that members of the tribe *Klebsielleae* are able to degrade 4-hydroxy-, and in some cases 3-hydroxybenzoate [271], that *Escherichia coli* is able to utilize 4-hydroxyphenylace-

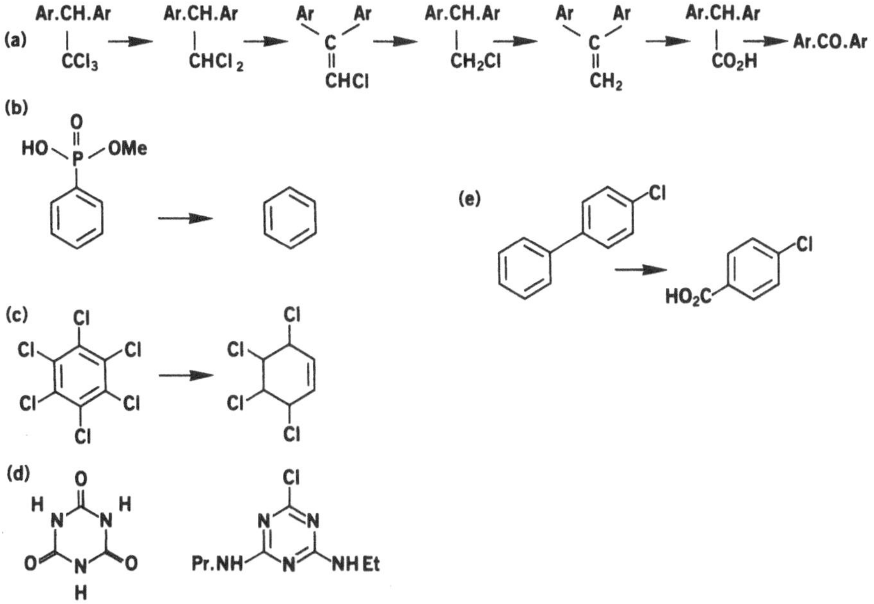

Fig. 29 a, b. Transformations (a–c, e) or growth substrates (d) involving facultatively anaerobic bacteria: (a) [274–276], (b) [234], (c) [277], (d) [278], (e) [103]

tate [272] and that several species of *Serratia* are able to degrade nicotinate [273]. Attention is particularly directed to some degradations carried out by facultative anaerobes (Fig. 29) which would be unusual for any organism. The most surprising is the transformation of phenyl methyl phosphonate into benzene. Although, the mechanism of this remains obscure, reference has already been made to formally comparable reductive eliminations (Fig. 5 a, b). Because a wide range of enteric bacteria may be found in the natural environment including many newly described taxa [279], it is timely to reconsider their role as agents of biodegradation. This is emphasized by the fact that some of the transformations (e.g. Fig. 29 a, c, d) occurred *only* under anaerobic conditions.

4. The metabolic potential of the anoxic phototrophic bacteria is well documented. Nonetheless, in spite of the demonstration of their ability to catabolize benzoate by a pathway involving reduction to cyclohexane carboxylate (Fig. 27) and to utilize methanol [280], few studies have attempted to assess their role in carbon metabolism in natural systems.

5. Anaerobic metabolism of amino acids and purines has been known for many years and has contributed greatly to our understanding of anaerobic metabolism. Anaerobic transformations, in general, have more recently attracted increasing interest because of the prevalence of anaerobic environments in natural situations. Attention is also briefly drawn to the existence of anaerobic fungi [281, 282] as a reminder of hitherto unexplored niches for eukaryotic microorganisms. Presumably these fungi are able to generate sufficient energy for growth by glycolysis. Therefore, the possible significance of such organisms in sediments containing fermentable substrates should not be forgotten.

The range of transformations carried out by anaerobic bacteria has widened considerably, and the following are especially notable:

– Dechlorination of γ-hexachlorocyclohexane [277, 283, 284] is of special interest considering the established persistence of this compound and the total ignorance of its fate under anaerobic conditions. The possibly underestimated frequency of such dechlorinations is also emphasized by the demonstration of the degradation of aromatic chloro carboxylates [285, 286]. This is especially significant because degradation was not apparently hindered by the existence of several chlorine atoms in the ring. It has also been shown that a consortium of bacteria including methanogens degraded simple aliphatic compounds such as chloroform and tetrachloroethane in the presence of acetate [287].
– It has been shown that enrichment with aromatic methoxy carboxylates such as vanillate and ferulate yields cultures of *Acetobacterium woodii* [288]. This organism brings about demethylation of the methoxy groups and produces acetate, but apparently leaves the aromatic ring intact.
– A new genus of anaerobic bacteria, *Pelobacter* [289], has been isolated after enrichment with a variety of polyhydroxylated aromatic carboxylates. Apparently, these organisms have an extremely restricted capacity to use other simpler substrates such as catechol.
– The metabolic versatility of anaerobic, sulphate-reducing bacteria is beginning to be realized. For example, the degradation of acetate [290], nicotinate [291], fatty acids up to C_{16} [291, 292], and furfural [293] under anaerobic conditions using sulphate as electron acceptor, are not only intrinsically interesting, but may be of considerable environmental significance in marine sediments where sulphate is abundant.
– Attention is drawn to anaerobic bacteria able to utilize oxalate [294] and succinate [295]. Although these substrates are readily degraded aerobically, considerable theoretical interest is attached to the existence of organisms which are able to make use of the small changes in free energy involved in the conversion of oxalate to formate, and succinate to propionate.

6. The heterotrophic potential of algae is well documented [296, 297]. In most studies, however, a limited range of carbohydrates, aliphatic carboxylic acids, amino acids, and glycerol has been examined. The ability of a range of algae to photometabolize a range of aromatic compounds such as naphthalene [298, 299], biphenyl [300], and aniline [301] suggests, however, an underestimation of their metabolic role in natural systems. It appears that algae are not able to use these compounds as sole sources of carbon and energy. Nevertheless this possibility clearly merits investigation because of the capacity of fungi and yeasts to bring about cleavage of the simpler aromatic rings [89, 257–259].

7. Protozoa have not been generally considered as important agents of biodegradation possibly because of their supposed nutritional fastidiousness. It has been clearly shown, however, that *Prototheca zopfii* is able to degrade a range of hydrocarbons [302, 303]. Furthermore, a syntrophic role for ciliates is suggested by recent work [304]. Some particularly arresting transformations carried out by *Tetrahymena thermophila* should also be noted: cell suspensions transformed pentachloronitrobenzene into a number of products including both the corresponding aniline and, more surprisingly, pentachlorothioanisole [305]. Although the

mechanism whereby the latter is produced is unknown, it presumably involved the intervention of glutathione. A similar transformation has also been demonstrated with fungi [306].

Plasmids

Plasmids carrying genes for antibiotic resistance [307] and resistance to heavy metals [308, 309] have been demonstrated in a wide range of bacteria. The importance of plasmids in enteric bacteria is shown, for example, by their roles in determining fermentation of carbohydrates [310–312], utilization of citrate [313] and decomposition of urea. All of these are relatively simple compounds. It has become increasingly clear, however, that plasmids carrying genes for the catabolism of a wide range of relatively recalcitrant substances are widely distributed. Some examples are given in Table 7. Moreover, the significance of these plasmids in natural ecosystems has not been fully resolved [334, 335]. Because few systematic studies on the natural occurrence of plasmids in non-clinical bacteria have been carried out [336], the frequency of their occurrence and their role in natural catabolism are indeed largely unknown. A further complication arises from the effect of temperature. In enteric bacteria, for example, thermosensitive plasmids which carry genes for utilization of citrate and for resistance to antibiotics and heavy metals have been demonstrated [337]. These plasmids have negligible rates of transmission at 37 °C but readily do so at 23 °C. It has therefore been argued plausibly that these plasmids may play an important role in the natural transmission of antibiotic resistance. Whether such a possibility may be realized for catabolic plasmids has not been resolved. It should also be noted that even if transmission to another organism takes place, phenotypic expression may not necessarily occur [338]. These and other mechanisms for genetic interactions have been discussed [335, 339]. It has been pointed out that certain experiments purporting to have evolved strains of bacteria capable of degrading 2,4,5-trichlorophenoxyacetate by plasmid transfer among a community of micro-organisms, might also plausibly be interpreted as involving selection during the lengthy incubation period. Indeed the cautious attitude adopted in these reviews and the care of the authors to distinguish between the *potential* and the *extent* of the operation of genetic transfer in nature is highly commendable.

Table 7. Examples of catabolic plasmids. References are given in brackets after the substrates

Octane [315]	Toluene [319]
Geraniol [316]	Xylene [320]
Camphor [317]	Naphthalene [321, 322, 323]
6-aminohexanoate dimer [318]	
Chloroacetate [324]	Salicylate [329]
4-chlorobiphenyl [213]	Parathion [330]
3-chlorobenzoate [325, 326]	Chloridazon [215]
Chlorophenoxyacetate [327, 328]	Nicotine/nicotinate [331, 332]
	Opines [333]

Attention is drawn lastly to the occurrence of plasmid-like DNA in eukaryotic yeasts and fungi [refs. in 340], even though the function of these has not been clearly resolved and hitherto there is no evidence supporting their role in catabolism.

Regulation

Regulation of catabolism is a matter of considerable significance for two reasons:

1. Organisms in their natural environment may never have been previously exposed to the test substance, or may have been exposed to it only intermittently.
2. Many other possible substrates will normally be present, and these may repress synthesis of the appropriate catabolic enzymes.

The general mechanisms whereby cells regulate their metabolism and the means whereby co-ordination of the metabolic pathways is achieved are well understood [341].

On the other hand, the catabolism of foreign substances has been much less extensively explored. However, details of the regulation of dissimilatory pathways particularly for aromatic compounds in pseudomonads have been established [342–345]. It is to be hoped that, with increasing interest in the genetics of other organisms such as the methylotrophic bacteria [131, 346, 347], a better overview of regulation in a wider range of organisms will become available. Until then, extrapolation from existing studies to other groups of organism must be made with caution. Certain subtle differences between bacteria and fungi must be pointed out. For example, while it appears that both groups metabolize catechols by identical pathways involving the *ortho* cleavage pathway, they differ in the operation of the protocatechuate (3,4-dihydroxybenzoate) pathway [348]. Bacteria produce 4-carboxymuconolactone and require two steps for its conversion into 3-ketoadipate. By contrast, fungal metabolism involves synthesis of 3-carboxymuconolactone which is converted by a single enzymatic step into 3-ketoadipate (Fig. 30).

Furthermore, the regulation of catabolic pathways is important in the design of experiments on biodegradability. The outcome of these experiments may be critically dependent on the choice of the substrate used for growth of the test organism. Many examples of this could be cited, but one will suffice: transformation of dibenzo-p-dioxin by *Pseudomonas putida* could be accomplished by cells grown with salicylate but not by those grown with succinate [349]. This is entirely consistent with the established regulation of the enzymes for catabolism of aromatic rings in this organism. For this reason, in studies on the degradation of aromatic compounds it has been convenient to use cultures grown with, for example, 4-hydroxybenzoate. An example of the possible complexity of catabolic pathways is the existence of both glutathione-dependent and glutathione-independent pathways in the degradation of 2,5-dihydroxybenzoates by bacilli [230].

In addition, attention should be drawn to three important considerations:

1. In some organisms the catabolic enzymes are constitutive, so that such organisms may enjoy a particular advantage in natural situations. On the other hand, their growth constitutes a considerable waste of genetic material.

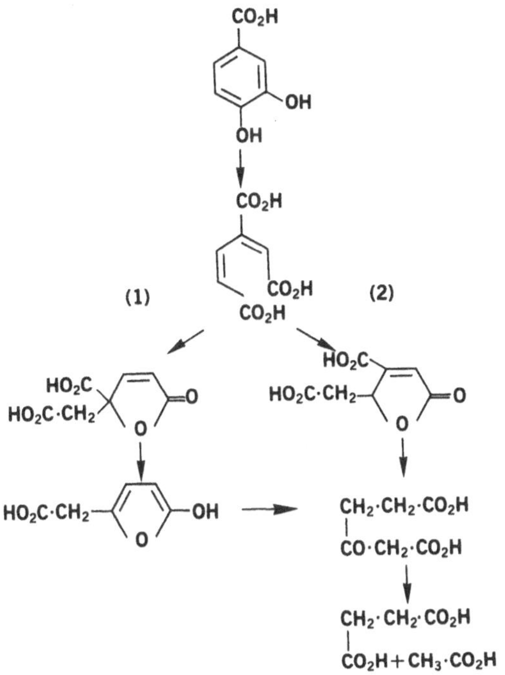

Fig. 30. Metabolism of 3,4-dihydroxybenzoate (protocatechuate) by (1) bacteria and (2) fungi [348]

2. In organisms carrying catabolic genes on plasmids, it has been clearly shown that effective degradation of a substrate may obligately involve both plasmid-borne and chromosomal genes (e.g. for degradation of n-alkanes [350] and naphthalene [322]). However, the reasons for such duplication are presently unknown. Both chromosomal- and plasmid-coded transport systems have been described in *Escherichia coli* and are apparently not closely related [351].
3. It appears that under nutrient limitation which frequently occurs in natural situations, there is a substantially weaker operation of catabolite repression. Under such circumstances, therefore, it may be possible for an organism to utilize two carbon substrates simultaneously [166].

Co-Metabolism: Co-Culture – Mixed Substrates: Mixed Cultures

In most of the studies discussed previously, only one type of organism and only a single substrate has been available. Such situations will seldom or never be encountered in natural situations so that attention has justifiably been given to examining metabolic processes in which more than one substrate is present. Nevertheless, it must be conceded that virtually all of the transformations carried out by a consortium of species, or by a single species in the presence of a co-substrate, have also been demonstrated for a single species in the presence of a single substrate. Unfortunately, a great deal of confusion has been generated resulting from investigations employing several substrates. The terms co-metabolism [15] and

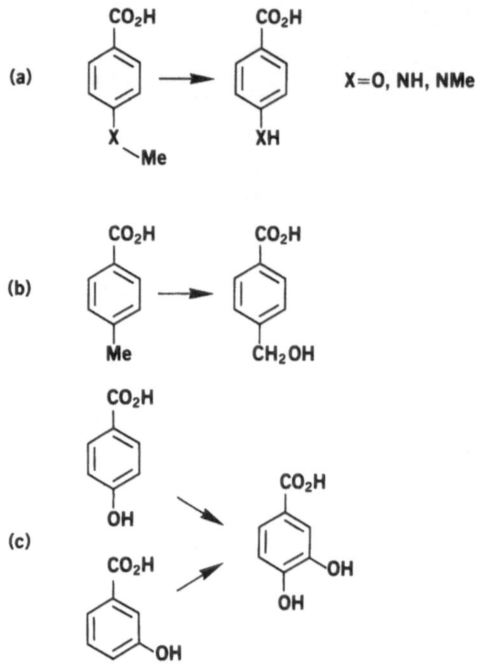

Fig. 31 a–c. Reactions catalysed by 4-methoxybenzoate demethylase [354]

co-oxidation [19] have been used, sometimes interchangeably. On the one hand, plausible and well-reasoned arguments have been given for rejecting the term co-metabolism as unnecessary [352], whereas on the other hand, reasonable arguments have also been made for the importance of the concept particularly in natural situations [15, 353]. It should be emphasized that the matter of regulation of catabolic pathways is a separate issue which has already been discussed. Leaving aside semantic controversies and metabolic hairsplitting, it seems clear, however, that important transformations are carried out taking advantage of two or more substrates.

Two very general points will be made:
1. Application of the term co-metabolism to situations in which only one substrate is present is a solecism, and such situations already appear to be adequately covered by the term transformation.
2. Extensive literature documents the capability of methylotrophic bacteria to carry out a very wide range of oxidations other than the oxidation of methane (Fig. 4). Although this appears to stem from the low substrate specificity of methane mono-oxygenase, it should be noted that 4-methoxybenzoate-O-demethylase also has a very low substrate specificity (Fig. 31) and that other oxygenases may also share this wide substrate versatility.

In reviewing experiments of the type noted in 2. above, it is hard to escape the conclusion that comparable results might have been obtained by incubating cell suspensions of the relevant organisms with the test substrate, a procedure hallowed by years of use. It is therefore reasonable to question what specific metabolic event in co-metabolism distinguishes it from other metabolic processes [352].

Nonetheless, situations in which two or more substrates are present are both frequent and important in the natural environment. Rather than take the presumptuous step of trying to amend the definition of co-metabolism which would result in further confusion, we tentatively suggest a new term-*concurrent metabolism*. Concurrent metabolism is the situation in which two substances are simultaneously degraded or transformed. The presence of two substrates is not obligatory, the substrate may not necessarily be transformed stoichiometrically into a single metabolite, and the additional substrate (co-substrate) may not increase either the rate or the yield of the transformation. We suggest that, in natural situations, concurrent metabolism may play a role at least as significant as that of co-metabolism.

Without undue prejudice, a number of comments are presented on experiments in which two substrates are simultaneously present.

1. Experiments on concurrent metabolism employing, for example, glucose as an additional substrate [110, 355, 356] seem to be of minimal environmental relevance because of the negligible concentrations of glucose in natural situations. Indeed, its presence may be inhibitory [357] and, on the most general grounds, its presence would not be expected to result in the induction of enzymes involved in catabolism of aromatic compounds. The essential role of glucose may be to increase the cell density of the culture. For example, experiments on the metabolism of fluorobenzoates [112], and of O,O-dimethyl-O-(3-methyl-4-nitrophenyl) phosphorothioate and 2,4-dichlorophenoxyacetate in mixed cultures [358] were interpreted as showing the beneficial effect of glucose. The higher cell densities in such cultures could, however, provide an alternative explanation. From an environmental point of view, the reduced lag before onset of degradation is of slight significance.

2. In experiments using two substrates, only partial degradation of one may occur because of the accumulation of toxic metabolites (e.g. 3-chlorocatechol from chlorobenzene [148]). Indeed this principle has been applied [359, 360] to the enrichment of mutants defective in catabolic genes. Additionally, organisms capable of utilizing 5-chlorosalicylate [95], 2,6-dichlorotoluene [7] and pentachlorophenol [91] were apparently unable to degrade non-halogenated substrates. The implication [361–363] that enhanced rates of degradation of halogenated pollutants could be achieved by deliberate addition of nonhalogenated analogues should, therefore, be viewed with caution.

3. While the presence of an unsubstituted analogue may enhance biodegradability e.g. 3,4-dichloroaniline [362, 363], its presence may, in other cases, be inhibitory. For example, degradation of 4-nitrobenzoate by a pseudomonad was competetively inhibited by benzoate, though metabolism of the latter was not affected by 4-nitrobenzoate [364].

4. Attention has already been very briefly drawn to the possibility that, under conditions of low substrate concentration, regulation of catabolic enzymes may be altered so that simultaneous utilization of more than one substrate occurs [166].

Although no general rules can be proposed, it is hoped that some of the drawbacks inherent in the design of experiments in biodegradation employing more than one substrate have emerged. Caution should therefore be exercised in contending that such experiments have greater environmental significance than those utilizing only single substrates.

Associations between organisms are obviously important and may be exceedingly complex. Some associations are so tight that the presence of more than one organism remained undetected over many years. Classic examples are *"Methanobacterium omelianskii"* [365] and *"Chloropseudomonas ethylica"* [366]. A more recent example is *"Methylobacterium ethanolicum"* [52]. These results emphasize the necessity for determining the purity of the organisms under study, and also suggest the probable ubiquity of such tight associations.

The first of the examples above illustrates a general phenomenon particularly prevalent among anaerobic bacteria, namely inter-organism hydrogen transfer [367–369]. In such situations, the metabolic products of the co-culture are markedly different from those of the individual organisms. Many examples have been provided. Although most of these have involved rumen bacteria, such interactions might also be important in certain types of anaerobic sediments. It may also be noted that, although most studies of inter-organism hydrogen transfer have been carried out with bacteria, it has recently been shown that stable co-cultures may be maintained between anaerobic, cellulose-degrading fungi and methanogenic bacteria [281]. A general review emphasizing anaerobic degradation of natural polymers such as starch, cellulose, pectin, and lignin under a variety of experimental conditions has been given by Zeikus [370]. This review nicely illustrates the importance of associations between organisms.

A mechanistic basis for the stability and functioning of mixed cultures clearly must exist. Although it is probably premature to produce a definitive rationalization, the following broad classification may be useful:

1. One of the organisms may provide an essential growth factor for the primary degrading organism. Some examples involving various vitamins have already been given (Table 1). In some cases, however, the nature of the necessary factor has not yet been resolved [79].

2. One of the organisms may produce lytic enzymes (proteases, nucleases, lipases) which enables it to utilize lysis products of one of the other organisms [371, 372].

3. One of the organisms serves as a scavenger of a metabolite which is toxic to the organism producing it. This may be a rather common situation and may be exemplified by the *Pseudomonas stutzeri-Pseudomonas aeruginosa* association involved in degradation of parathion [44] or the *Pseudomonas-Hyphomicrobium* association in the assimilation of methane [373]. It appears that inter-organism hydrogen transfer among anaerobes serves to reduce the hydrogen tension of sensitive organisms and simultaneously achieves a balance in the redox state of the system. Further details and examples are given in reviews cited above.

4. The presence of both organisms may be metabolically essential to the success of the degradation. A few examples are given:

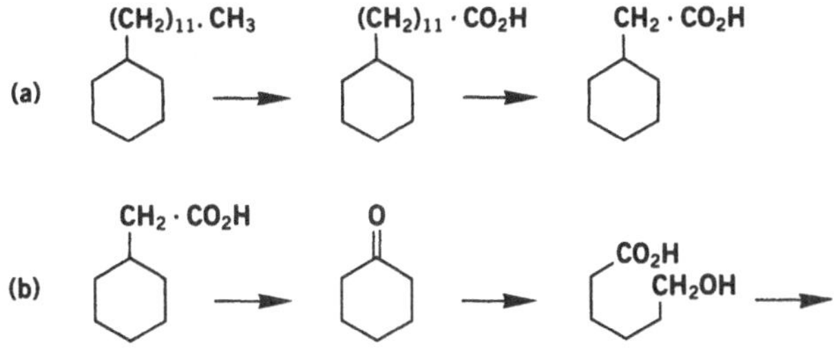

(a) $(CH_2)_{11} \cdot CH_3$ ⬡ → $(CH_2)_{11} \cdot CO_2H$ ⬡ → $CH_2 \cdot CO_2H$ ⬡

(b) $CH_2 \cdot CO_2H$ ⬡ → O⬡ → CO_2H, CH_2OH ⬡ →

Fig. 32 a, b. Metabolism of dodecylcyclohexane by a consortium of bacteria (a) *Mycobacterium rhodochrous* (*Rhodococcus rhodochrous*) and (b) *Arthrobacter* sp. [376]

- Anaerobic β-oxidation of carboxylic acids carried out by *Syntrophomonas* [374] is dependent on the presence of other organisms capable of using the reduced equivalents.
- A mixed microflora was able to degrade 2-chlorobiphenyl with the production of 2-chlorobenzoate which was resistant for further microbial attack [375]. Other organisms are therefore necessary to complete the mineralization of the original substrate. However, 2-chlorobenzoate can hardly be considered a recalcitrant substrate.
- By β-oxidation, alkyl cyclohexanes having an even number of carbon atoms produce cyclohexane acetate which is not further metabolized by the organism but can be totally degraded by addition of another organism (Fig. 32). In our own studies, we have encountered a comparable situation in the formation of 2-hydroxynaphthalene-3-carboxylate during metabolism of anthracene [377]. This is not further degraded by the organism producing it but can be degraded by other organisms. It would be plausible to suggest that the same would apply to 2-formyl-3-hydroxy-benzothiophene produced from dibenzothiophene [378] which is isoelectronic with anthracene.

It should, however, be noted that single strains may be able to accomplish all of the reactions. For example, strains of the genus *Desulfococcus* [291] and *Desulfonema* [292] can carry out β-oxidation of carboxylates in pure cultures. It is clearly premature to speculate on the relative significance of mixed culture transformations as opposed to those carried out by single organisms. Because of the probably greater frequency of the former conditions in natural situations, their potential importance should not, however, be underestimated.

Environmental Significance

In the previous sections, various aspects of the metabolism of recalcitrant xenobiotics have been presented. It is now necessary to discuss their possible consequences to the ecosystem. The following discussion is divided between: the effects presented from the point of view of the organism carrying out the metabolism and the effects presented, from the wider perspective of the ecosystem as a whole.

The Viewpoint of the Organism

Two situations may be envisaged:
1. The substrate is non-toxic to the organism, but a toxic metabolite is produced
2. The organism has developed a detoxification mechanism
   – for the initial substrate and/or
   – for a metabolite produced.
1. The production of toxic metabolites which inhibit further transformation of the original substrate is widely encountered. The classic example is the toxicity of fluoroacetate: fluorocitrate produced through operation of the initial stages of the tricarboxylic acid cycle, inhibits aconitase so that continued functioning of the cycle is impossible [379]. It is interesting to note, that fluoroacetate formed during the metabolism of 2-fluoro-4-nitrobenzoate by *Nocardia erythropolis* could be converted by cell-free extracts into fluorocitrate [380]. Relevant examples may also be drawn from studies on the metabolism of aromatic compounds. For example, (a) dihydroxydibenzodioxin produced from dibenzodioxin [349] and (b) 3-chlorocatechol produced during "co-metabolism" of chlorobenzene and succinate [148] or as an intermediate in the degradation of 3-chlorobenzoate [9], inhibit the oxygenases required for further metabolism. Organisms may, however, have evolved catabolic pathways to avoid the synthesis of toxic metabolites: this has been discussed in a previous section. The end result is the total degradation of the substrate. For the sake of completeness it should be pointed out that the synthesis of toxic metabolites is not restricted to microbial systems. Comparable situations are encountered in mammalian systems, for example, where aromatic hydrocarbons are metabolized to carcinogenic epoxides before their conjugation and eventual excretion.
2. A substantial number of compounds are relatively toxic to microorganisms. For example, nitro- and chlorophenols are toxic to a wide range of aerobic organisms because of their ability to uncouple oxidative phosphorylation. Microorganisms have therefore evolved a variety of devices to detoxify such substances. These generally involve relatively simple chemical modifications of the structure.
(i) For toxic substrates, three broad classes of reaction may be distinguished:

Esterification

Acetylation: e.g. phenols [150], aromatic amines [225, 381], chloramphenicol [382]
Phosphorylation: e.g. streptomycin [307]
Sulphate formation: e.g. phenols [253, 254]

Alkylation

C-alkylation: e.g. Hg^{2+} [308, 309, 383]
O-alkylation: e.g. chlorophenols and chloroguaiacols [82, 144, 150–152], oximes [156]
N-alkylation: e.g. ribosomal RNA [158]

It must be emphasized that C-methylation of metals and the other O- and N-methylations have little in common. Methylation of Hg^{2+} plausibly involves a coenzyme-B_{12}-mediated reaction involving formally Me^- [383]. Although the mechanisms of the other methylations have not been examined systematically, in all cases the methyl group probably originates from S-adenosyl methionine [384]. The activity of a catechol-O-methyltransferase might plausibly be invoked for the phenols, although this enzyme has not been shown to produce veratroles [155].

Nucleophilic Displacements

Reactions involving displacement of halogen, nitro, and sulphate substituents do not generally occur (Fig. 5c, d). The formation of thioanisoles from pentachloronitrobenzene [305, 306] does, however, belong to this class of reaction and presumably occurs via reactions involving glutathione. Although this is a relatively neglected transformation, it might also possibly occur with polyhalogenated aromatic compounds such as hexachlorobenzene or polychlorinated biphenyls. Both fungi [385] and bacteria [386] have been shown to utilize aromatic thiomethyl groups as sulphur sources, so that further transformations of such compounds may be accomplished through the action of microbial consortia.

(ii) Detoxification may also be necessary to deal with toxic metabolites. Two simple examples may be given. Organisms capable of degrading aliphatic hydrocarbons are frequently sensitive to the carboxylic acids [387] which are inevitably produced during degradation of the substrate. The formation of esters formed by reaction between these carboxylic acids and the corresponding alcohols (which are their precursors) has been observed [388]. Indeed, in the cases of 1-chlorohexadecane, the ester was virtually the only product produced by *Micrococcus cerificans* (*Acinetobacter calcoaceticus*) [389]. Chloroalkanes may also be incorporated into cell lipids by comparable reactions involving ω-oxidation with or without alteration in the length of the alkyl side-chain [390].

Complex transformations have been demonstrated with aromatic nitro compounds and result in a variety of products including azo- and azoxy-compounds and, from *ortho* nitroanilines, benziminazoles (Fig. 24).

The Viewpoint of the Ecosystem

So far as the ecosystem in general is concerned, the significance of any metabolic activity cannot be assessed only from the point of view of the organism producing the metabolite. In environmental hazard assessments of a substance, possibly insufficient attention has been directed to the situation in which a metabolite, though non-toxic to the organism producing it, is highly toxic to other components of the system. Examples are the production of extra-cellular toxins by bacteria, of mycotoxins by fungi and of antibiotics as products of secondary metabolism by both fungi and bacteria [391]. However, it has been less generally recognized that metabolites from xenobiotics could be toxic to other organisms in the ecosystem.

Three examples will be discussed briefly in order to illustrate the general principles involved.

1. The first example is one which aroused widespread concern over environmental issues. Microbial C-methylation is one of the detoxification systems developed by an organism in response to exposure to Hg^{2+}. This results in the synthesis of compounds which are less toxic than Hg^{2+} to the microorganisms but which are extremely toxic to higher forms of life and have a high potential for accumulation in fatty tissues. Discharge of Hg^{2+} may also have a more subtle effect in favoring organisms capable of bringing about methylation and selecting against those organisms unable to do so. Hg-methylation capacity is widely distributed in the microbial world and has been demonstrated in the intestinal contents of fish [392] and humans [393]. All of these considerations make it difficult to assess the distribution and long-term consequences of these processes. They clearly merit continuing investigation.
2. Microbial nitrosation of dimethylamine has been demonstrated [394] although the picture is complicated by purely abiological reactions between nitrite and the secondary amine. Although the N-nitroso compound is apparently innocuous to the organisms producing it, such compounds are mutagenic in a number of test systems and probably carcinogenic to mammals [395].
3. Although the formation of O-methyl ethers from halogenated phenolic compounds has been documented [81], the toxicity of these products has not been previously evaluated. The possible environmental consequences have been the subject of a recent investigation from this laboratory and may be briefly summarized:
   - The chloroveratroles showed a high potential for accumulation in laboratory fish. The bioconcentration potential of the other neutral metabolites was inferred from values for the relative mobility on C_{18} HPTLC plates.
   - In a test on their effect on embryos and larvae of zebra fish, the metabolites were as toxic as their precursors (chlorophenols and chloroguaiacols). In addition, they induced serious curvature of the larvae and deformation of the notochord. Neither of these defects had been observed with the non-methylated compounds. Because a number of other effects, such as haemorrhage in muscle tissue and gross distortion of the abdomen, were noted only at high concentrations, their significance is difficult to evaluate quantitatively.
   - Both 3,4,5-trichloro- and tetrachloro-veratroles were found in the liver fat of fish caught in areas into which chloroguaiacols are discharged.

On the basis of these findings, it is concluded that a hazard assessment of the chlorophenols and chloroguaiacols based solely on data for these compounds, and which failed to take into consideration the properties of their metabolites, could be seriously misleading.

Although the natural occurrence of O-methylation has never been systematically examined, accumulating data suggest that it is widely distributed. This provides a plausible explanation for the detection of pentachloroanisole in oysters [396] and in the muscle tissue of fish [397] following an accidental spill of pentachlorophenol, and of di-O-methyl-tetrabromobisphenol-A in sediment [398] and mussels [399]. The biological effects of the neutral bromo compounds do not appear to have been examined.

In none of the cases cited has an alternative source of the *O*-methyl compounds been traced. Although *O*-methylation may plausibly be regarded as an effective detoxification system for the micro-organisms, it might seriously disrupt the functioning of other biota in the ecosystem.

The significant role of experiments on biodegradability and biotransformation was pointed out at the beginning of this review. The results presented here clearly demonstrate the need for caution in interpreting the results of such experiments. It may not be sufficient to investigate only the degradation (or transformation) of the substance being examined. Attention must also be specifically directed to the possibility that metabolites produced from the original substrate may be at least as toxic. This toxicity must be evaluated not only for the organism carrying out the metabolism but also for other components of the natural ecosystem.

A valid environmental hazard assessment requires that a number of additional factors be taken into consideration. Among these, the following are of greatest importance:
1. The occurrence of the relevant transformation under natural conditions.
2. The demonstration that biological effects observed in the laboratory occur also in natural wild populations.
3. The distribution and fate of the compound or its transformation products in fish, water or sediment from areas subjected to discharge of the substance under consideration.

It is hoped that some of the difficulties in implementing, and the pitfalls in interpreting experiments on biodegradation of recalcitrant compounds, have emerged from this review.

Summary

An account is given of principles and methods applicable to the study of the biodegradation and biotransformation of recalcitrant compounds. Only the salient issues are summarized:
1. The basic procedure involves incubation of cell suspensions of microorganisms with the substrate to be examined. Periodic analysis for the concentrations of the substrate and possible metabolites is than carried out. The organisms used are obtained by enrichment of natural samples of soil, sediment or water.
2. Enrichment procedures are discussed and attention directed to the following: continuous culture methods, the source of the inoculum, problems of water insolubility and toxicity, and methods for isolating the required organisms.
3. Analytical methods are briefly discussed, and attention is drawn to problems with substrates having high vapor pressures and to difficulties arising from binding of the substrate to cells and to the walls of the containing vessels.
4. A scheme of metabolic patterns is presented, and emphasis is placed on the importance of transient, toxic, or lipophilic metabolites.
5. Physical and chemical parameters influencing biodegradation are discussed, whereby the importance of the concentration and bioavailability of the substrate is emphasized.

6. The value of mechanistic considerations is stressed. Particular attention is directed to the effect of substituents (halogen, nitro, and sulphonate) and to differences in the metabolic pathways used by bacteria and fungi.
7. Examples are given of degradations and transformations occurring under anaerobic conditions both in the presence and absence of electron acceptors other than oxygen. The possibly underestimated role of algae and protozoa is pointed out.
8. The occurrence and importance of catabolic plasmids are noted, and the significance of regulatory mechanisms is examined, followed by a discussion of metabolism when more than one substrate, or more than one organism is present.
9. The environmental significance of metabolic processes is evaluated both from the viewpoint of the metabolizing organism and from that of the ecosystem as a whole.
10. A brief discussion of the relevance of these studies to environmental hazard assessment is presented.

Acknowledgement

It is a pleasure to thank Lars Landner for supporting our studies, for stimulating discussions over the years, and for a critical appraisal of this review which has removed at least its more glaring errors and its major obscurities.

References

1. Landner, L.: *In* Proceedings of the international MARTOX Symposium, Ghent 1983. Eds. Persoon, G. et al. In press
2. Alexander, M.: Microbial Ecol. *2*, 17 (1975)
3. Fenical, W.: Science *215*, 923 (1982)
4. Gerike, P., Fischer, W.K.: Ecotoxicol. Environ. Safety *5*, 45 (1981)
5. Bourquin, A.W., Przybyszewski, V.A.: Appl. Environ. Microbiol. *34*, 411 (1977)
6. Franklin, F.C.H., Bagdasarian, M., Timmis, K.N.: *In* Microbial degradation of xenobiotics and recalcitrant compounds. FEMS Symposium No. 12. Eds. Leisinger T. et al. p. 109. Academic Press, London (1981)
7. Vandenbergh, P.A., Olsen R.H., Colaruotolo, J.F.: Appl. Environ. Microbiol. *42*, 737 (1981)
8. Chakrabarty, A.M.: *In* Biodegradation and detoxification of environmental pollutants. Ed. Chakrabarty, A.M., p. 127. CRC Press, Boca Raton (1982)
9. Reineke, W. et al.: J. Bacteriol. *150*, 195 (1982)
10. Schwien, U., Schmidt, E.: Appl. Environ. Microbiol. *44*, 33 (1982)
11. Clarke, P.H.: Proc. R. Soc. Lond. B *207*, 385 (1980)
12. Mortlock, R.P.: Ann. Rev. Microbiol. *36*, 259 (1982)
13. Fukui, S., Tanaka, A.: Ann. Rev. Microbiol. *36*, 145 (1982)
14. Postgate, J.R.: The fundamentals of nitrogen fixation. Cambridge University Press, Cambridge (1982)
15. Horvath, R.S.: Bact. Rev. *36*, 146 (1972)
16. Dagley, S.: Essays Biochem. *11*, 81 (1975)
17. Dagley, S.: Surv. Prog. Chem. *8*, 121 (1977)
18. Dagley, S.: Quart. Rev. Biophys. *11*, 577 (1978)
19. Perry, J.J.: Microbiol. Rev. *43*, 59 (1979)
20. Slater, J.H., Sommerville, H.J.: Symp. Soc. Gen. Microbiol. *22*, 221 (1979)
21. Smith, R.V., Davis, P.J.: Adv. Biochem. Eng. *14*, 61 (1980)

22. Alexander, M.: Science *211*, 132 (1981)
23. Atlas, R.M.: Microbiol. Rev. *45*, 180 (1981)
24. Kobayashi, H., Rittman, B.E.: Environ. Sci. Technol. *16*, 170 A (1982)
25. Lal, R., Saxena, D.M.: Microbiol. Rev. *46*, 95 (1982)
26. Wood, J.M.: Environ. Sci. Technol. *16*, 291 A (1982)
27. Quayle, J.R., Bull, A.T.: Phil. Trans. R. Soc. Lond. B *297*, 445 (1982)
28. Hill, I.R., Wright, S.J.L.: Pesticide microbiology. Academic Press, London (1978)
29. Gibson, D.T.: Microbial degradation of organic compounds. Marcel Dekker, New York (1984)
30. Leisinger, T. et al.: Microbial degradation of xenobiotics and recalcitrant compounds. FEMS Symp. No. 12. Academic Press, London (1981)
31. Shewan, J.M., McMeekin, T.A.: Ann. Rev. Microbiol. *37*, 233 (1983)
32. Van Niel, C.B.: J. Gen. Microbiol. *13*, 201 (1955)
33. Palleroni, N.J.: In The prokaryotes. Eds. Starr, M.P. et al. Vol. 1, p. 655. Springer-Verlag, Berlin (1981)
34. Winogradsky, S.: Microbiologie du sol. Oevres complètes. Masson et Cie, Paris (1949)
35. Beijerinck, M.W.: Verzamelde Geschriften, vol. 1–5. Nijhof, den Haag (1921)
36. Horikoshi, K., Akiba, T.: Alkalophilic microorganisms: A new microbial world. Springer-Verlag, Berlin (1982)
37. Kuznetsov, S.I., Dubinina, G.A., Lapteva, M.A.: Ann. Rev. Microbiol. *33*, 377 (1977)
38. Poindexter, J.S.: Adv. Microbiol. Ecol. *5*, 63 (1981)
39. Jannasch, H.W., Wirsen, C.O.: Appl. Environ. Microbiol. *43*, 1116 (1982)
40. Yayanos, A.A., Dietz, A.S., Van Boxtel, R.: Appl. Environ. Microbiol. *44*, 1356 (1982)
41. Tabor, P.S. et al.: Appl. Environ. Microbiol. *44*, 413 (1982)
42. Claus, D., Walker, N.: J. Gen. Microbiol. *36*, 107 (1964)
43. Harris, R., Knowles, C.J.: J. Gen. Microbiol. *129*, 1005 (1983)
44. Daughton, C.G., Hsieh, D.P.H.: Appl. Environ. Microbiol. *34*, 175 (1977)
45. Harder, W.: In Microbial degradation of xenobiotics and recalcitrant compounds. FEMS Symposium No. 12. Eds. Leisinger, T. et al. p. 77. Academic Press, London (1981)
46. Harder, W., Dijkhuizen, L.: Phil. Trans. R. Soc. London B *297*, 459 (1982)
47. Parkes, R.J.: In Microbial interactions and communities. Eds. Bull, A.T., Slater, J.H. p. 45. Academic Press, London (1982)
48. Palleroni, N.J., Doudoroff, M.: Ann. Rev. Phytopath. *10*, 73 (1972)
49. Van Veen, W.L.: Antonie van Leeuwenhoek. J. Microbiol. Serol. *39*, 189 (1973)
50. Jensen, H.L.: Can. J. Microbiol. *3*, 151 (1957)
51. Miura, Y. et al.: J. Ferment. Technol. *56*, 339 (1978)
52. Lidstrom-O'Connor, M.E., Fulton, G.L., Wopat, A.E.: J. Gen. Microbiol. *129*, 3139 (1983)
53. Moore, J.K., Braymer, H.D., Larson, A.D.: Appl. Environ. Microbiol. *46*, 316 (1983)
54. Knowles, R.: Microbiol. Rev. *46*, 43 (1982)
55. Aranha, H.G., Brown, L.R.: Appl. Environ. Microbiol. *42*, 74 (1981)
56. Sylvestre, M. et al.: Appl. Environ. Microbiol. *44*, 871 (1982)
57. Minard, R.D., Russel, S., Bollag, J.-M.: J. Agric. Food Chem. *25*, 841 (1977)
58. Corke, C.T. et al.: J. Agric. Food Chem. *27*, 644 (1979)
59. Bordeleau, L.M., Rosen, J.D., Bartha, R.: J. Agric. Food Chem. *20*, 573 (1972)
60. Miller, J.M., Gray, D.O.: J. Gen. Microbiol. *128*, 1803 (1982)
61. Focht, D.D., Williams, F.D.: Can. J. Microbiol. *16*, 309 (1970)
62. De Bont, J.A.M., Van Dijken, J.P., Harder, W.: J. Gen. Microbiol. *127*, 315 (1981)
63. Stafford, D.A., Callely, A.G.: J. Gen. Microbiol. *55*, 285 (1969)
64. Stapley, E.O., Starkey, R.L.: J. Gen. Microbiol. *64*, 77 (1970)
65. Cook, A.M., Daughton, C.G., Alexander, M.: J. Bacteriol. *133*, 85 (1978)
66. Neidhardt, F.C., Bloch, P.L., Smith, D.F.: J. Bacteriol. *119*, 736 (1974)
67. Leadbetter, E.R., Foster, J.W.: Arch. Microbiol. *30*, 91 (1958)
68. Baruah, J.N., Alroy, Y., Mateles, R.I.: Appl. Microbiol. *15*, 961 (1967)
69. Sylvestre, M.: Appl. Environ. Microbiol. *39*, 1223 (1980)
70. Kiyohara, H., Nagao, K., Yana, K.: Appl. Environ. Microbiol. *43*, 454 (1982)
71. Shiaris, M.P., Cooney, J.J.: Appl. Environ. Microbiol. *45*, 706 (1983)

72. Doudoroff, M.: Enzymologia 9, 59 (1940)
73. Robert-Gero, M., Poiret, M., Stanier, R.Y.: J. Gen. Microbiol. 57, 207 (1969)
74. Merkel, G.J., Underwood, W.H., Perry, J.J.: FEMS Microbiol. Lett. 3, 81 (1978)
75. Goodfellow, M., Williams, S.T.: Ann. Rev. Microbiol. 37, 189 (1983)
76. Hirsch, C.F., Christensen, D.L.: Appl. Environ. Microbiol. 46, 925 (1983)
77. Mackay, S.J.: Appl. Environ. Microbiol. 33, 227 (1977)
78. Songer, J.G.: Can. J. Microbiol. 27, 1 (1981)
79. Sakazawa, C. et al.: Appl. Environ. Microbiol. 41, 261 (1981)
80. Phillips, W.E., Perry, J.J.: Int. J. System. Bacteriol. 26, 220 (1976)
81. Neilson, A.H. et al.: Can. J. Fish. Aquat. Sci. 41, 1502 (1984)
82. Neilson, A.H. et al.: Appl. Environ. Microbiol. 45, 774 (1983)
83. Barry, A.L., Thornsberry, C.: In Manual of clinical microbiology. 3rd. ed. p. 463. Eds. Lenette, E.H., Balows, A., Hausler, W.J. American Society for Microbiology, Washington D.C. (1980)
84. Perdue, E.M., Wolfe, N.L.: Environ. Sci. Technol. 16, 847 (1982)
85. Blau, K., King, G.S. (Eds.): Handbook of derivatives for chromatography. Heyden, London (1978)
86. Knapp, D.R. Handbook of analytical derivatization reactions. Wiley, New York (1979)
87. Pierce, A.E.: Silylation of organic compounds. Pierce Chemical Company, Rockford (1979)
88. Chu, J.P., Kirsch, E.J.: Appl. Microbiol. 23, 1033 (1972)
89. Walker, N.: Soil Biol. Biochem. 5, 525 (1973)
90. Tyler, J.E., Finn, R.K.: Appl. Microbiol. 28, 181 (1974)
91. Stanlake, G.J., Finn, R.K.: Appl. Environ. Microbiol. 44, 1421 (1982)
92. Karns, J.S. et al.: Appl. Environ. Microbiol. 46, 1176 (1983)
93. Hughes, D.E.: Biochem. J. 96, 181 (1965)
94. Walker, N., Harris, D.: Soil Biol. Biochem. 2, 27 (1970)
95. Crawford, R.L., Olson, P.E., Frick, T.D.: Appl. Environ. Microbiol. 38, 379 (1979)
96. Hartmann, J., Reineke, W., Knackmuss, H.-J.: Appl. Environ. Microbiol. 37, 421 (1979)
97. Klages, U., Lingens, F.: FEMS Microbiol. Lett. 6, 201 (1979)
98. Klages, U., Lingens, F.: Z. Bakt. Parasitenk. Infect. Hyg. Abt. 1. Orig. C 1, 215 (1980)
99. Klages, U., Markus, A., Lingens, F.: J. Bacteriol. 146, 64 (1981)
100. Walker, N., Wiltshire, G.H.: J. Gen. Microbiol. 12, 478 (1955)
101. Ahmed, M., Focht, D.D.: Can. J. Microbiol. 19, 47 (1973)
102. Furukawa, K., Chakrabarty, A.M.: Appl. Environ. Microbiol. 44, 619 (1982)
103. Sylvestre, M., Fauteux, J.: J. Gen. Appl. Microbiol. (Tokyo) 28, 61 (1982)
104. Gibson, D.T. et al.: Biochemistry 7, 3795 (1968)
105. Morris, C.M., Barnsley, E.A.: Can. J. Microbiol. 28, 73 (1982)
106. Klecka, G.M., Gibson, D.T.: Appl. Environ. Microbiol. 39, 288 (1980)
107. Spokes, J.R., Walker, N.: Arch. Microbiol. 96, 125 (1974)
108. Knackmuss, H.-J., Hellwig, M.: Arch. Microbiol. 117, 1 (1978)
109. Engelhardt, G., Rast, H.G., Wallnöfer, P.R.: Arch. Microbiol. 114, 25 (1977)
110. Rosenberg, A., Alexander, M.: J. Agric. Food Chem. 28, 297 (1980)
111. Horvath, R.S.: J. Agric. Food Chem. 19, 291 (1971)
112. Horvath, R.S., Flathman, P.: Appl. Environ. Microbiol. 31, 889 (1976)
113. Clarke, K.F. et al.: Biochim. Biophys. Acta 404, 169 (1975)
114. Schreiber, A. et al.: Appl. Environ. Microbiol. 39, 58 (1980)
115. Gray, H.H.P., Thornton, H.G.: C. Bakt. Parasitenk. Infect. Hyg. 73, 74 (1928)
116. Murphy, J.F., Stone, R.W.: Can. J. Microbiol. 1, 579 (1955)
117. Davies, J.I., Evans, W.C.: Biochem. J. 91, 251 (1964)
118. Evans, W.C., Fernley, H.N., Griffiths, E.: Biochem. J. 95, 819 (1965)
119. Barnsley, E.A.: Biochem. Biophys. Res. Comm. 72, 1116 (1976)
120. Kiyohara, H., Nagao, K.: J. Gen. Microbiol. 105, 69 (1978)
121. Patel, T.R., Barnsley, E.A.: J. Bacteriol. 143, 668 (1980)
122. Ensley, B.D., Gibson, D.T., Laborde, A.L.: J. Bacteriol. 149, 948 (1982)
123. Ribbons, D.W., Eaton, R.W.: In Biodegradation and detoxification of environmental pollutants. Ed. Chakrabarty, A.M. p. 59. CRC Press, Boca Raton (1982)

124. Barnsley, E.A.: J. Bacteriol. *154*, 113 (1983)
125. Levering, P.R. et al.: Arch. Microbiol. *129*, 72 (1981)
126. Green, P.N., Bousefield, I.J.: Int. J. System. Bacteriol. *33*, 875 (1983)
127. Colby, J., Dalton, H., Whittenbury, R.: Ann. Rev. Microbiol. *33*, 481 (1979)
128. Higgins, I.J. et al.: Microbiol. Rev. *45*, 556 (1981)
129. Whittenbury, R., Dalton, H.: *In* The prokaryotes. Eds. Starr, M.P. et al. Vol. 1, p. 894. Springer-Verlag, Berlin (1981)
130. Anthony, C.: The biochemistry of methylotrophs. Academic Press, London (1982)
131. Haber, C.L. et al.: Science *221*, 1147 (1983)
132. Colby, J., Stirling, D.I., Dalton, H.: Biochem. J. *165*, 395 (1977)
133. Patel, R.N. et al.: Appl. Environ. Microbiol. *44*, 1130 (1982)
134. Higgins, I.J. et al.: Biochem. Biophys. Res. Comm. *89*, 671 (1979)
135. Patt, T.E., Cole, G.C., Hanson, R.S.: Int. J. System. Bacteriol. *26*, 226 (1976)
136. Patel, R.N., Hou, C.T., Felix, A.: J. Bacteriol. *136*, 352 (1978)
137. Lynch, M.L., Wopat, A.E., O'Connor, M.L.: Appl. Environ. Microbiol. *40*, 400 (1980)
138. Wolf, H.J., Hanson, R.S.: J. Gen. Microbiol. *114*, 187 (1979)
139. Furukawa, K., Matsumura, F.: J. Agric. Food. Chem. *24*, 251 (1976)
140. Furukawa, K., Tomizuka, N., Kamibayashi, A.: Appl. Environ. Microbiol. *38*, 301 (1979)
141. De Frenne, E., Eberspächer, J., Lingens, F.: Eur. J. Biochem. *33*, 357 (1973)
142. Evans, W.C. et al.: Biochem. J. *122*, 543 (1971)
143. Johnston, H.W., Briggs, G.G., Alexander, M.: Soil Biol. Biochem. *4*, 187 (1972)
144. Suzuki, T.: J. Pesticide Sci. *8*, 419 (1983)
145. Steenson, T.I., Walker, N.: J. Gen. Microbiol. *16*, 146 (1957)
146. Loos, M.A., Roberts, R.N., Alexander, M.: Can. J. Microbiol. *13*, 691 (1967)
147. Raymond, D.G.M., Alexander, M.: Pest. Biochem. Physiol. *1*, 123 (1971)
148. Klecka, G.M., Gibson, D.T.: Appl. Environ. Microbiol. *41*, 1159 (1981)
149. Sariaslani, F.S., Rosazza, J.P.: Appl. Environ. Microbiol. *46*, 468 (1983)
150. Rott, B., Nitz, S., Korte, F.: J. Agric. Food, Chem. *27*, 306 (1979)
151. Cserjesi, A.J., Johnson, E.L.: Can. J. Microbiol. *18*, 45 (1972)
152. Gee, J.M., Peel, J.L.: J. Gen. Microbiol. *85*, 237 (1974)
153. Guldberg, H.C., Marsden, C.A.: Pharmacol. Rev. *27*, 135 (1975)
154. Müller-Enoch, D. et al.: Z. Naturforsch. *31*c, 509 (1976)
155. Veser, J., Geywitz, P., Thomas, H.: Z. Naturforsch. *34*c, 709 (1979)
156. Harper, D.B., Nelson, J.: J. Gen. Microbiol. *128*, 1667 (1982)
157. Tanaka, T., Weisblum, B.: J. Bacteriol. *123*, 771 (1975)
158. Skinner, R.H., Cundcliffe, E.: J. Gen. Microbiol. *128*, 2411 (1982)
159. Steinbüchel, A. et al.: J. Gen. Microbiol. *129*, 2825 (1983)
160. Rubin, H.E., Subba-Rao, R.V., Alexander, M.: Appl. Environ. Microbiol. *43*, 1133 (1982)
161. Subba-Rao, R.V., Rubin, H.E., Alexander, M.: Appl. Environ. Microbiol. *43*, 1139 (1982)
162. Boethling, R.S., Alexander, M.: Appl. Environ. Microbiol. *37*, 1211 (1979)
163. Van der Kooij, D., Visser, A., Hijnen, W.A.M.: Appl. Environ. Microbiol. *39*, 1198 (1980)
164. Van der Kooij, D., Oranje, J.P., Hijnen, W.A.M.: Appl. Environ. Microbiol. *44*, 1086 (1982)
165. Van der Kooij, D., Hijnen, A.M.: Appl. Environ. Microbiol. *45*, 804 (1983)
166. Harder, W., Dijkhuizen, L.: Ann. Rev. Microbiol. *37*, 1 (1983)
167. Khan, S.U.: Res. Rev. *52*, 1 (1974)
168. Parris, G.E.: Res. Rev. *76*, 1 (1980)
169. Sjoblad, R.D., Bollag, J.-M.: Appl. Environ. Microbiol. *33*, 906 (1977)
170. Liao, W. et al.: Environ. Sci. Technol. *16*, 403 (1982)
171. Parris, G.E.: Environ. Sci. Technol. *14*, 1099 (1980)
172. Bollag, J.-M., Minard, R.D., Liu, S.-Y.: Environ. Sci. Technol. *17*, 72 (1983)
173. Gjessing, E.T., Berglund, L.: Vatten *38*, 402 (1982)
174. Gjessing, E.T., Berglund, L.: Arch. Hydrobiol. *21*, 24 (1981)

175. Rast, H.G. et al.: FEMS Microbiol. Lett. *8*, 259 (1980)
176. Subba-Rao, R.V., Alexander, M.: Appl. Environ. Microbiol. *44*, 659 (1982)
177. Macdonald, C.R., Copper, D.G., Zajic, J.E.: Appl. Environ. Microbiol. *41*, 117 (1981)
178. Kaplan, N., Rosenberg, E.: Appl. Environ. Microbiol. *44*, 1335 (1982)
179. Suzuki, T. et al.: Agric. Biol. Chem. *33*, 1619 (1969)
180. Kretschmer, A., Bock, H., Wagner, F.: Appl. Environ. Microbiol. *44*, 864 (1982)
181. Itoh, S., Suzuki, T.: Agric. Biol. Chem. *36*, 2233 (1972)
182. Shoham, Y., Rosenberg, E.: J. Bacteriol. *156*, 161 (1983)
183. Cooper, D.G., Paddock, D.A.: Appl. Environ. Microbiol. *46*, 1426 (1983)
184. Hansch, C., Leo, A.: Substituent constant for correlation analysis in chemistry and biology. Wiley, New York (1979)
185. Paris, D.F., Wolfe, N.L., Steen, W.C.: Appl. Environ. Microbiol. *44*, 153 (1982)
186. Kunz, D.A., Weimer, P.J.: J. Bacteriol. *156*, 567 (1983)
187. Wegener, W.S. et al.: Bact. Rev. *32*, 1 (1968)
188. Omori, T., Alexander, M.: Appl. Environ. Microbiol. *35*, 867 (1978)
189. Allison, N., Skinner, A.J., Cooper, R.A.: J. Gen. Microbiol. *129*, 1283 (1983)
190. Schaeffer, T.I. et al.: Appl. Environ. Microbiol. *38*, 742 (1979)
191. Fall, R.R., Brown, J.L., Schaeffer, T.I.: Appl. Environ. Microbiol. *38*, 715 (1979)
192. Stirling, L.A., Watkinson, R.J., Higgins, I.J.: J. Gen. Microbiol. *99*, 119 (1977)
193. Anderson, M.S., Hall, R.A., Griffin, M.: J. Gen. Microbiol. *120*, 89 (1980)
194. Gallo, M. et al.: Biochim. Biophys. Acta *296*, 624 (1973)
195. Gunsalus, I.C., Peterson, T.C., Sligar, S.G.: Ann. Rev. Biochem. *44*, 377 (1975)
196. Cardini, G., Jurtshuk, P.: J. Biol. Chem. *245*, 2789 (1970)
197. Yamada, Y. et al.: Appl. Microbiol. *29*, 400 (1975)
198. De Bont, J.A.M., Peck, M.W.: Arch. Microbiol. *127*, 99 (1980)
199. Catterall, F.A., Murray, K., Williams, P.A.: Biochim. Biophys. Acta *237*, 361 (1971)
200. Jerina, D.M. et al.: Arch. Biochem. Biophys. *142*, 394 (1971)
201. Reiner, A.M., Hegeman, G.D.: Biochemistry *10*, 2530 (1971)
202. Patel, T.R., Gibson, D.T.: J. Bacteriol. *119*, 879 (1974)
203. Jefrey, A.M. et al.: Biochemistry *14*, 575 (1975)
204. Ferris, J.P. et al.: Arch. Biochem. Biophys. *156*, 97 (1973)
205. Gerniglia, C.E. et al.: Arch. Microbiol. *117*, 135 (1978)
206. Daly, J.W., Jerina, D.M., Witkop, B.: Experientia *28*, 1129 (1972)
207. Faulkner, J.K., Woodcock, D.: J. Chem. Soc. 1187 (1965)
208. Guroff, G., Kondo, K., Daly, J.: Biochem. Biophys. Res. Commun. *25*, 622 (1966)
209. Dalton, H. et al.: J. Chem. Soc. Chem. Comm. 482 (1981)
210. Knackmuss, H.-J. et al.: Zbl. Bakt. Hyg. I. Abt. Orig. B *162*, 127 (1976)
211. Leidner, H. et al.: Xenobiotica *10*, 47 (1980)
212. Kulla, H.G. et al.: Arch. Microbiol. *135*, 1 (1983)
213. Furukawa, K.: *In* Biodegradation and detoxification of environmental pollutants. Ed. Chakrabarty, A.M. p.33. CRC Press, Boca Raton (1982)
214. Dagley, S., Johnson, P.A.: Biochim. Biophys. Acta *78*, 377 (1963)
215. Kreis, M., Eberspächer, J., Lingens, F.: Zbl. Bakt. Hyg., I. Abt. Orig. C *2*, 45 (1981)
216. Goldman, P., Milne, G.W.A., Pignatoro, M.T.: Arch. Biochem. Biophys. *118*, 178 (1967)
217. Engesser, K.-H., Schmidt, E., Knackmuss, H.-J.: Appl. Environ. Microbiol. *39*, 68 (1980)
218. Cain, R.B., Farr, D.R.: Biochem. J. *106*, 859 (1968)
219. Brilon, C., Beckmann, W., Knackmuss, H.-J.: Appl. Environ. Microbiol. *42*, 44 (1981)
220. Cartwright, N.J., Cain, R.B.: Biochem. J. *71*, 248 (1959)
221. McCormick, N.G., Cornell, J.H., Kaplan, A.M.: Appl. Environ. Microbiol. *35*, 945 (1978)
222. Corbett, M.D., Corbett, B.R.: Appl. Environ. Microbiol. *41*, 942 (1981)
223. Hallas, L.E., Alexander, M.: Appl. Environ. Microbiol. *45*, 1234 (1983)
224. Laanio, T.L., Kearney, P.C., Kaufman, D.D.: Pest. Biochem. Physiol. *3*, 271 (1973)
225. Kaufman, D.D., Plimmer, J.R., Klingebiel, U.I.: J. Agric. Food Chem. *21*, 127 (1973)
226. Callely, A.G.: Prog. Ind. Microbiol. *14*, 205 (1978)
227. Sutherland, J.B., Crawford, D.L., Pometto, A.L.: Appl. Environ. Microbiol. *41*, 442 (1981)

228. Ander, P., Eriksson, K.-E., Yu, H.: Arch. Microbiol. *136*, 1 (1983)
229. Pometto, A.L. III, Sutherland, J.B., Crawford, D.L.: Can. J. Microbiol. *27*, 636 (1981)
230. Crawford, R.L., Frick, T.D.: Appl. Environ. Microbiol. *34*, 170 (1977)
231. Hirschberg, R., Ensign, J.C.: J. Bacteriol. *108*, 757 (1971)
232. Trudgill, P.W.: Biochem. J. *113*, 577 (1969)
233. Watson, G.K., Cain, R.B.: Biochem. J. *146*, 157 (1975)
234. Cook, A.M., Daughton, C.G., Alexander, M.: Biochem. J. *184*, 453 (1979)
235. Dutton, P.L., Evans, W.C.: Biochem. J. *113*, 525 (1969)
236. Guyer, M., Hegeman, G.: J. Bacteriol. *99*, 906 (1969)
237. Hutber, G.N., Ribbons, D.W.: J. Gen. Microbiol. *129*, 2413 (1983)
238. Williams, R.J., Evans, W.C.: Biochem. J. *148*, 1 (1975)
239. Taylor, B.F., Hearn, W.L., Pincus, S.: Arch. Microbiol. *122*, 301 (1979)
240. Aftring, R.P., Chalker, B.E., Taylor, B.F.: Appl. Environ. Microbiol. *41*, 1177 (1981)
241. Benckiser, G., Ottow, J.C.G.: Appl. Environ. Microbiol. *44*, 576 (1982)
242. Smith, D.I., Callely, A.G.: J. Gen. Microbiol. *91*, 210 (1975)
243. Taylor, D.G., Trudgill, P.W.: J. Bacteriol. *134*, 401 (1978)
244. Blakely, E.R., Papish, B.: Can. J. Microbiol. *28*, 1324 (1982)
245. Kilbane, J.J. et al.: Appl. Environ. Microbiol. *44*, 72 (1982)
246. Cerniglia, C.E., Gibson, D.T.: Arch. Biochem. Biophys. *186*, 121 (1978)
247. Cerniglia, C.E.: J. Gen. Microbiol. *128*, 2055 (1982)
248. Schwartz, R.D., Williams, A.L., Hutchinson, D.B.: Appl. Environ. Microbiol. *39*, 702 (1980)
249. Smith, R.V. et al.: Biochem. J. *196*, 369 (1981)
250. Cerniglia, C.E., Fu, P.P., Yang, S.K.: Appl. Environ. Microbiol. *44*, 682 (1982)
251. Wong, L.K. et al.: Appl. Environ. Microbiol. *46*, 1239 (1983)
252. Cerniglia, C.E., Crow, S.A.: Arch. Microbiol. *129*, 9 (1981)
253. Cerniglia, C.E., Freeman, J.P., Mitchum, R.K.: Appl. Environ. Microbiol. *43*, 1070 (1982)
254. Golbeck, J.H., Albaugh, S.A., Radmer, R.: J. Bacteriol. *156*, 49 (1983)
255. Ferris, J.P. et al.: Arch. Biochem.Biophys. *175*, 443 (1976)
256. Smith, R.V., Rosazza, J.P.: Arch. Biochem. Biophys. *161*, 551 (1974)
257. Anderson, J.J., Dagley, S.: J. Bacteriol. *146*, 219 (1981)
258. Gaal, A., Neujahr, H.Y.: Arch. Microbiol. *130*, 54 (1981)
259. Oka, T. et al.: Can. J. Microbiol. *17*, 111 (1971)
260. Baumann, P. et al.: Int. J. System. Bacteriol. *33*, 857 (1983)
261. Strøm, A.R., Olafsen, J.A., Larsen, H.: J. Gen. Microbiol. *112*, 315 (1979)
262. Takagi, M., Tsuchiya, T., Ishimoto, M.: J. Bacteriol. *148*, 762 (1981)
263. Kwan, H.S., Barrett, E.L.: J. Bacteriol. *155*, 1147 (1983)
264. Kapralek, F.: J. Gen. Microbiol. *115*, 133 (1972)
265. Tuttle, J.H., Jannasch, H.W.: J. Gen. Microbiol. *115*, 732 (1973)
266. Yen, H.-C., Marrs, B.: Arch. Biochem. Biophys. *181*, 411 (1977)
267. Zinder, S.H., Brock, T.D.: Arch. Microbiol. *116*, 35 (1978)
268. Taylor, B.F., Ribbons, D.W.: Appl. Environ. Microbiol. *46*, 1276 (1983)
269. Taylor, B.F.: Appl. Environ. Microbiol. *46*, 1286 (1983)
270. Lütgens, M., Gottschalk, G.: J. Gen. Microbiol. *128*, 1915 (1982)
271. Moscoso-Vizarra, M., Popoff, M.: Ann. Microbiol. *128* A, 199 (1977)
272. Burlingame, R., Chapman, P.J.: J. Bacteriol. *155*, 113 (1983)
273. Grimond, P.A.D. et al.: J. Gen. Microbiol. *98*, 39 (1977)
274. Mendel, J.L., Walton, M.S.: Science *151*, 1527 (1966)
275. Wedemeyer, G.: Appl. Microbiol. *15*, 569 (1967)
276. Baarschers, W.H., Bharath, A.I., Elvish, J.: Can. J. Microbiol. *28*, 176 (1982)
277. Jagnow, G., Haider, K., Ellwardt, P.-C.: Arch. Microbiol. *115*, 285 (1977)
278. Jessee, J.A. et al.: Appl. Environ. Microbiol. *45*, 97 (1983)
279. Ferragut, C. et al.: Int. J. System. Bacteriol. *33*, 133 (1983)
280. Sahm, H., Cox, R.B., Quayle, J.R.: J. Gen. Microbiol. *94*, 313 (1976)
281. Mountfort, D.O., Asher, R.A., Bauchop, T.: Appl. Environ. Microbiol. *44*, 128 (1982)
282. Natvig, D.O., Gleason, F.H.: Arch. Microbiol. *134*, 5 (1983)

283. Ohisa, N., Yamaguchi, M., Kurihara, N.: Arch. Microbiol. *125*, 221 (1980)
284. Ohisa, N., Kurihara, N., Nakajima, M.: Arch. Microbiol. *131*, 330 (1982)
285. Suflita, J.M. et al.: Science *218*, 1115 (1982)
286. Horowitz, A., Suflita, J.M., Tiedje, J.M.: Appl. Environ. Microbiol. *45*, 1459 (1983)
287. Bouwer, E.J., McCarty, P.L.: Appl. Environ. Microbiol. *45*, 1286 (1983)
288. Bache, R., Pfennig, N.: Arch. Microbiol. *130*, 255 (1981)
289. Schink, B., Pfennig, N.: Arch. Microbiol. *133*, 195 (1982)
290. Brandis-Heep, A. et al.: Arch. Microbiol. *136*, 222 (1983)
291. Imhoff-Stuckle, D., Pfennig, N.: Arch. Microbiol. *136*, 194 (1983)
292. Widdel, F., Kohring, G.-W., Mayer, F.: Arch. Microbiol. *134*, 286 (1983)
293. Brune, G., Schoberth, S.M., Sahm, H.: Appl. Environ. Microbiol. *46*, 1187 (1983)
294. Dawson, K.A., Allison, M.J., Hartman, P.A.: Appl. Environ. Microbiol. *40*, 833 (1980)
295. Schink, B., Pfennig, N.: Arch. Microbiol. *133*, 209 (1982)
296. Droop, M.R.: *In* Algal physiology and biochemistry. Ed. Stewart, W.D.P. p. 530. Blackwell Scientific Publications, Oxford (1974)
297. Neilson, A.H., Lewin, R.A.: Phycologia *13*, 227 (1974)
298. Cerniglia, C.E., Gibson, D.T., Van Baalen, C.: J. Gen. Microbiol. *116*, 495 (1980)
299. Cerniglia, C.E. et al.: J. Gen. Microbiol. *128*, 987 (1982)
300. Cerniglia, C.E., Van Baalen, C., Gibson, D.T.: Arch. Microbiol. *125*, 203 (1980)
301. Cerniglia, C.E., Freeman, J.P., Van Baalen, C.: Arch. Microbiol. *130*, 272 (1981)
302. Walker, J.D., Pore, R.S.: Appl. Environ. Microbiol. *35*, 694 (1978)
303. Koenig, D.W., Ward, H.B.: Appl. Environ. Microbiol. *45*, 333 (1983)
304. Rogerson, A., Berger, J.: J. Gen. Appl. Microbiol. *29*, 41 (1983)
305. Murphy, S.E., Drotar, A., Fall, R.: Chemosphere *11*, 33 (1982)
306. Nakanishi, T., Oku, H.: Phytopathol. *59*, 59 (1969)
307. Foster, T.J.: Microbiol. Rev. *47*, 361 (1983)
308. Summers,A.O., Silver, S.: Ann. Rev. Microbiol. *32*, 637 (1978)
309. Silver, S., Kinschef, T.G.: *In* Biodegradation and detoxification of environmental pollutants. Ed. Chakrabarty, A.M. p. 85. CRC Press Inc.: Boca Raon (1982)
310. Ørskov, I., Ørskov, F.: J. Gen. Microbiol. *77*, 487 (1973)
311. Smith, H.W., Parsell, Z.: J. Gen. Microbiol. *87*, 129 (1975)
312. Le Minor, L., Coynault, C.: Ann. Microbiol. *127* A, 214 (1976)
313. Ishiguro, N. et al.: J. Bacteriol. *149*, 961 (1982)
314. Chakrabarty, A.M.: Ann. Rev. Genet. *10*, 7 (1976)
315. Chakrabarty, A.M., Chou, G., Gunsalus, I.C.: Proc. Nat. Acad. Sci. USA *70*, 1137 (1973)
316. Vandenbergh, P.H., Wright, A.M.: Appl. Environ. Microbiol. *45*, 1953 (1983)
317. Rheinwald, J.G., Chakarabarty, A.M., Gunsalus, I.C.: Proc. Nat. Acad. Sci. USA *70*, 885 (1973)
318. Negoro, S. et al.: J. Bacteriol. *143*, 238 (1980)
319. Worsey, M.J. et al.: J. Bacteriol *134*, 757 (1978)
320. Friello, D.A. et al.: J. Bacteriol. *127*, 1217 (1976)
321. Dunn, N.W., Gunsalus, I.C.: J. Bacteriol. *114*, 974 (1973)
322. Zuniga, M.C., Durham, D.R., Welch, R.A.: J. Bacteriol. *147*, 836 (1981)
323. Donnors, M.A., Barnsley, E.A.: J. Bacteriol. *149*, 1096 (1982)
324. Kawasaki, H., Yahara, H., Tonomura, K.: Agric. Biol. Chem. *45*, 1477 (1981)
325. Chaterjee, D.K. et al.: J. Bacteriol. *146*, 639 (1981)
326. Chaterjee, D.K., Chakrabarty, A.M.: Mol. Gen. Genet. *188*, 279 (1982)
327. Fisher, P.R., Appleton, J., Pemberton, J.M.: J. Bacteriol. *135*, 798 (1978)
328. Don, R.H., Pemberton, J.M.: J. Bacteriol. *145*, 681 (1981)
329. Chakrabarty, A.M.: J. Bacteriol. *112*, 815 (1972)
330. Serdar, C.M. et al.: Appl. Environ. Microbiol. *44*, 246 (1982)
331. Thacker, R. et al.: J. Bacteriol. *135*, 289 (1978)
332. Brandsch, R., Hinkkanen, A.E., Decker, K.: Arch. Microbiol. *132*, 26 (1982)
333. Nester, E.W., Konsuge, T.: Ann. Rev. Microbiol. *35*, 531 (1981)
334. Reanney, D.C., Roberts, W.P., Kelly, W.J.: *In* Microbial interactions and communities. Eds. Bull, A.T., Slater, J.H. p. 287. Academic Press, London (1982)

335. Williams, P.W.: Phil. Trans. R. Soc. Lond. B *297*, 631 (1982)
336. Burton, N.F., Day, M.J., Bull, A.T.: Appl. Environ. Microbiol. *44*, 1026 (1982)
337. Smith, H.W. et al.: J. Gen. Microbiol. *109*, 37 (1978)
338. Nakazawa, T. et al.: J. Bacteriol. *134*, 270 (1978)
339. Reanney, D.C., Gowland, P.C., Slater, J.H.: Symp. Soc. Gen. Microbiol. *34*, 379 (1983)
340. Stahl, U. et al.: Mol. Gen. Genet. *178*, 639 (1980)
341. Mandelstam, J., McQuillen, K., Dawes, I.: Biochemistry of microbial growth. 3rd. ed. Blackwell, Oxford (1982)
342. Ornston, L.N.: Bacteriol. Rev. *35*, 87 (1971)
343. Wheelis, M.: Ann. Rev. Microbiol. *29*, 505 (1975)
344. Cain, R.: *In* Lignin biodegradation: microbiology, chemistry and potential applications. Eds. Kirk, T.K., Higuchi, T., Chang, H. Vol. 1, p. 21. CRC Press, Boca Raton (1980)
345. Ornston, L.N., Yeh, W.K.: *In* Biodegradation and detoxification of environmental pollutants. Ed. Chakrabarty, A.M. p. 105. CRC Press, Boca Raton (1982)
346. Moore, A.T., Nayudu, M., Holloway, B.W.: J. Gen. Microbiol. *129*, 785 (1983)
347. Tatra, P.K., Goodwin, P.M.: J. Gen. Microbiol. *129*, 2629 (1983)
348. Cain, R.B., Bilton, R.F., Darrah, J.A.: Biochem. J. *108*, 797 (1968)
349. Klecka, G.M., Gibson, D.T.: Biochem. J. *180*, 639 (1979)
350. Grund, A. et al.: J. Bacteriol. *123*, 546 (1975)
351. Reynolds, C.H., Silver, S.: J. Bacteriol. *156*, 1019 (1983)
352. Hulbert, M.H., Krawiec, S.: J. Theor. Biol. *69*, 287 (1977)
353. Dalton, H., Stirling, D.I.: Phil. Trans. R. Soc. Lond. B *297*, 481 (1982)
354. Bernhardt, F.-H. et al.: Eur. J. Biochem. *35*, 126 (1973)
355. Shoda, M., Udaka, S.: Appl. Environ. Microbiol. *39*, 1129 (1980)
356. Schukat B. et al.: Curr. Microbiol. *9*, 81 (1983)
357. Rubin, H.E., Alexander, M.: Environ. Sci. Technol. *17*, 104 (1983)
358. Liu, D. et al.: Environ. Sci. Technol. *15*, 788 (1981)
359. Wigmore, G.J., Ribbons, D.W.: J. Bacteriol. *143*, 816 (1980)
360. Wigmore, G.J., Ribbons, D.W.: J. Bacteriol. *146*, 920 (1981)
361. Jacobson, S.N., Alexander, M.: Appl. Environ. Microbiol. *42*, 1062 (1981)
362. You, I.-S., Bartha, R.: Appl. Environ. Microbiol. *44*, 678 (1982)
363. You, I.-S., Bartha, R.: J. Agric. Food. Chem. *30*, 274 (1982)
364. Haller, H.D., Finn, R.K.: Appl. Environ. Microbiol. *35*, 890 (1978)
365. Bryant, W.P.: Arch. Microbiol. *59*, 20 (1967)
366. Olson, J.M.: Int. J. System. Bacteriol. *28*, 128 (1978)
367. Bauchop, T., Mountfort, D.O.: Appl. Environ. Microbiol. *42*, 1103 (1981)
368. Mah, R.A.: Phil. Trans. R. Soc. Lond. B *297*, 599 (1982)
369. Wolin, M.J.: *In* Microbial interactions and communities. Eds. Bull, A.T., Slater, J.H. Vol. 1, p. 323 (1982)
370. Zeikus, J.G.: Symp. Soc. Gen. Microbiol. *34*, 423 (1983)
371. Senior, E., Bull, A.T., Slater, J.H.: Nature (London) *263*, 476 (1976)
372. Linton, J.D., Buckee, J.C.: J. Gen. Microbiol. *101*, 219 (1977)
373. Wilkinson, T.G., Topiwala, H.H., Hammer, G.: Biotechnol. Bioeng. *16*, 41 (1974)
374. McInerny, M.J. et al.: Appl. Environ. Microbiol. *41*, 1029 (1981)
375. Shiaris, M.P., Sayler, G.S.: Environ. Sci. Technol. *16*, 367 (1982)
376. Feinberg, E.L., Ramage, P.I.N., Trudgill, P.W.: J. Gen. Microbiol. *121*, 507 (1980)
377. Rogoff, M.N., Wender, J.: J. Bacteriol. *74*, 108 (1957)
378. Laborde, A.L., Gibson, D.T.: Appl. Environ. Microbiol. *34*, 783 (1977)
379. Peters, R.A.: Proc. R. Soc. Lond. B *139*, 143 (1952)
380. Cain, R.B., Tranter, E.K., Darrah, J.A.: Biochem. J. *106*, 211 (1968)
381. Van Alfen, N.K., Kosuge, T.: J. Agric. Food Chem. *22*, 221 (1974)
382. Shaw, W.V., Brodsky, R.F.: J. Bacteriol. *95*, 28 (1968)
383. Wood, J.M., Wang, H.-K.: Environ. Sci. Technol. *17*, 582 A (1983)
384. Suzuki, T.: J. Pesticide Sci. *3*, 441 (1978)
385. Murray, D.S., Rieck, W.L., Lynd, J.Q.: Appl. Microbiol. *19*, 11 (1970)
386. Cook, A.M., Hüttner R.: Appl. Environ. Microbiol. *43*, 781 (1982)
387. Atlas, R.M., Bartha, R.: Antonie van Leeuwenhoek J. Microbiol. Serol. *39*, 257 (1973)

388. Finnerty, W.R., Hawtrey, E., Kallio, R.E.: Z. Allg. Microbiol. *2*, 169 (1962)
389. Kolattukudy, P.E., Hankin, L.: J. Gen. Microbiol. *54*, 145 (1968)
390. Murphy, G.L., Perry, J.J.: J. Bacteriol. *156*, 1158 (1983)
391. Leisinger, T., Margraff, R.: Microbiol. Rev. *43*, 422 (1979)
392. Rudd, J.W.M., Furutani, A., Turner, M.A.: Appl. Environ. Microbiol. *40*, 777 (1980)
393. Rowland, I.R., Grasso, P., Davies, M.J.: Experientia *31*, 1064 (1975)
394. Kunisaki, N., Hayashi, M.: Appl. Environ. Microbiol. *37*, 279 (1979)
395. Magee, P.N., Barnes, J.M.: Br. J. Cancer *10*, 114 (1956)
396. Miyazaki, T. et al.: Bull. Environ. Contam. Toxicol. *26*, 577 (1981)
397. Renberg, L. et al.: Ambio *12*, 121 (1983)
398. Watanabe, I., Kashimoto, T., Tatsukawa, R.: Bull. Environ. Contam. Toxicol. *31*, 48 (1983)
399. Watanabe, I., Kashimoto, T., Tatsukawa, R.: Chemosphere *12*, 1533 (1983)

Biodegradation of Water-Soluble Compounds

H. A. Painter, E. F. King

Water Research Centre
Stevenage SG1 1TH, United Kingdom

Introduction

Interest in biodegradability grew as a necessity when problems were created by the use and subsequent discharge of chemicals which did not degrade but persisted in the environment. Probably the best known example is the introduction of the first synthetic detergents in the 1950's. These detergents, now described as "hard," were only partially removed in sewage treatment and passed through treatment works, largely unchanged, to rivers. It was common, at least in the UK, to see huge banks of foam created by aeration in activated sludge plants and also at weirs and other sites of turbulence on polluted rivers. The problem, caused by the inability of aquatic bacteria to degrade sufficiently rapidly the highly branched chain alkyl benzene sulphonates, was satisfactorily solved by replacement by the manufacturers of the "hard" type with a biodegradable ("soft") product in which the degree of branching had been considerably reduced.

Since then, there have been no comparable widespread spectacular effects of non-biodegradable chemicals, probably because few other synthetic chemicals are used so widely in such large quantities as are synthetic detergents. However, there is concern over chemicals which, though not used in very large quantities, are known to be persistent and to accumulate in aquatic organisms. Some of these chemicals have been detected far from the initial sites of use. Pesticides, for example, have been detected in remote parts of the oceans; DDT for instance was reported to be found, at concentrations up to 2 µg/l, in the Sargasso Sea.

Chemicals discharged to the aquatic environment may cause damage in a number of ways. They can directly inhibit micro-organisms in sewage treatment processes and in rivers, thus possibly seriously interfering in aerobic treatment of waste waters and in the self-purification processes in bodies of water, and may kill aquatic organisms, e.g. fish. Adsorbed and insoluble chemicals may interfere in the anaerobic treatment of sewage sludge and, if subsequently applied to land as fertiliser, may inhibit essential microbial action in soil and become assimilated into plants. Animals and man could thus become affected by drinking polluted water and eating contaminated food. Hence, some form of scrutiny of chemicals is necessary before they are put into widespread use and in many countries this examination has been embodied in appropriate legislation.

It has long been recognised that micro-organisms, especially bacteria, are the main agents by which polluting organic matter (sewage, decaying vegetation, etc.) is removed in the aquatic environment so that a study of biodegradation properly includes an understanding of microbial nutrition, metabolism, and adaptation.

This chapter is confined to aerobic biodegradation of water-soluble chemicals in aquatic systems: the important topics – anaerobic biodegradation, degradation in soil, and insoluble chemicals – have not been so well investigated and no agreed methods of testing have yet been reached.

In order to make clear what is subsequently meant, the chapter begins with a glossary of terms, followed by a brief account of the nature of biodegradation. Next comes a discussion of important factors in testing and a description of the various test methods, with the emphasis on internationally agreed methods. There follow discussions on the choice of methods, schemes of testing, interpretation of results and legal requirements. Finally, a short section anticipates future research needs.

Glossary of Terms

Acclimatisation (or acclimation) includes those processes, such as selection and adaptation by which a mixed population of micro-organisms develops the ability to degrade a substance, hitherto not biodegradable. It also covers the situation in which the populations develop tolerances to inhibitory substances.

Activated sludge is the flocculated mixture of micro-organisms and inert organic and inorganic material produced by aeration of sewage and/or waste waters.

Biodegradability is the capacity of a substance to undergo microbial attack.

Biodegradation is the breakdown of a compound by micro-organisms and can be:
  (i) *Primary* – a change in the chemical structure of a substance resulting in loss of a specific property of that substance;
 (ii) *Environmentally acceptable* – degradation to such an extent as to remove undesirable properties of the compound. This frequently corresponds to primary biodegradation, but may vary depending on the circumstances under which the products are discharged to the environment;
(iii) *Ultimate* – the complete breakdown of a compound to fully oxidised (aerobically) simple molecules (e.g. CO_2, H_2O, NO_3^-, NH_4^+) and the formation of new cells.

Bioelimination is the removal of a compound from the liquid phase in the presence of living micro-organisms by physico-chemical as well as biological processes.

BOD (Biochemical oxygen demand) is the amount of oxygen consumed by micro-organisms when metabolising a compound.

COD (Chemical oxygen demand) is the amount of oxygen consumed during the oxidation of a compound with hot acid dichromate and provides a measure of the oxidisable matter present.

Co-metabolism is the process(es) by which a normally non-biodegradable compound is biodegraded only in the presence of an additional source of carbon.

Degradation is any process by which the structure of a compound is simplified.

Die-Away test is a batch test for biodegradability in which the decrease in the initial concentration of the compound is observed with time.

Inherently Biodegradable is a term applied to a classification of chemicals for which there is unequivocal evidence of primary or ultimate biodegradation in any test of biodegradability.

Inhibition is the effect(s) of a compound, or its metabolites, on microorganisms which may be manifested as a reduction in respiration rate, substrate degradation, gas evolution or growth.

Readily Biodegradable is a term applied to an arbitrary classification of chemicals which have passed specified screening tests for ultimate biodegradability. The tests are so stringent that such compounds biodegrade rapidly and completely in a wide variety of aerobic environments.

Screening tests are relatively simple, batch tests which may be used for preliminary assessment of biodegradability or toxicity of a test compound.

Simulation tests are designed to predict the rate of biodegradation of a compound under relevant environmental conditions.

Toxicity is the extent to which a test compound adversely affects microorganisms.

Treatability is the amenability of compounds to removal during biological treatment without adversely affecting the normal operation of the treatment processes.

Nature of Biodegradation

In the aquatic environment organic compounds are removed by a variety of mechanisms – biological, autoxidation, adsorption, sedimentation, hydrolysis, photodegradation – but it is generally accepted that biological processes play the major role. Of the many agents involved, bacteria are thought to be the most important in metabolising both natural and xenobiotic compounds. Bacteria are ubiquitous and their size, specific growth rates, metabolic versatility and mode of life make them eminently suitable for this function.

Very many single species of bacteria have been shown, in laboratory experiments, to be capable of degrading and growing at the expense of naturally-occurring compounds, present in media as the sole source of carbon, and of converting them to carbon dioxide, water, ammonia, etc.; that is, of completely mineralising organic compounds. Such growth and degradation reactions depend on a number of chemical and physical factors. Besides a source of carbon for synthesis of new cells and for energy (except autotrophic species), elements required in assimilable form by bacteria include nitrogen, phosphorus, sulphur, calcium, and magnesium in relatively high concentration and others such as iron, copper, manganese, and zinc in trace amounts. Some species also require pre-formed members of the B group of vitamins. The macro-elements carbon, nitrogen, and phosphorus should be present approximately in the ratio $50:5:1$ otherwise growth decreases or stops entirely.

An adequate supply of molecular oxygen must be available; for most aerobic species at least 1 mg/l should be present for maximal growth. Anaerobically, bound oxygen in the form of nitrate, carbonate or sulphate is acceptable. There are certain other factors such as temperature, pH value, salinity, osmotic pressure, concentration of toxic agents, which have ranges beyond which growth will not occur. All of these requirements and factors have to be borne in mind when assessing biodegradability. The biochemical and enzymological details of the degradation of organic compounds are outside the scope of this chapter but can be found throughout the biochemical literature [e.g. 1, 2].

However, such mono-cultures studied under laboratory conditions do not necessarily reflect what happens in the aquatic environment, where mixed microbial populations exist and develop as complex communities in the presence of large numbers of substances both natural and man-made. The study of such communities, even in the laboratory, is only just beginning and very little is yet known about the interplay of individual species bringing about the degradation of a given substance in the environment or of the metabolic pathways which are used.

Although a large number of diverse biodegradative mechanisms exists, many compounds, especially xenobiotic and anthropogenic compounds, are reported to be undegraded under circumstances apparently adequate for microbial growth. These reports raise doubts about the earlier belief, expressed for example by Gale [3], that all organic compounds could be eventually biodegraded (principle of microbial infallibility) and Alexander [4] has suggested that some compounds are indeed non-biodegradable or recalcitrant. This problem will not be readily resolved, since man-made chemicals have been in the environment for only a relatively short period of time compared with that during which evolution is thought to have occurred of the now well-known microbial catabolic sequences by which naturally occurring compounds are broken down.

Recalcitrance

There can be simple reasons for non-biodegradation in a laboratory experiment and these must be discounted before embarking on an explanation of recalcitrance of a given chemical. The simplest explanation is that the species present did not possess the metabolic capability – they were not competent – although elsewhere such species may exist. The chemical may not have been available to the micro-organisms, because of insolubility or adsorption on to suspended matter, or the chemical may have been inhibitory to micro-organisms at the concentration used. Finally, the physico-chemical conditions may have been inappropriate.

At the enzymic level, the compound may not be degraded because it is unable to enter the cell. Although the necessary enzymes may be present to degrade the substance within the cell, the organism may not possess the specific transport systems to deal with the new chemical.

A chemical may be recalcitrant because the organism does not possess the appropriate enzyme or series of enzymes necessary to convert the compound to intermediates of a known, central metabolic pathway. Even minor changes in the structure of readily degradable compounds usually produce analogues which, because of the high degree of specificity of the original enzymes, are recalcitrant.

Another reason for recalcitrance is the inability of a compound to induce the required catabolic enzymes, although suitable enzymes, inducible by other substrates, may be present.

Co-Metabolism

Biodegradation has so far been considered as a growth process in which some of the organic compound has been oxidised, the resulting energy being used to synthesise new cells from the remainder of the substrate. A new phenomenon [5] is recognised in which an organism grows at the expense of one substrate, but also has the ability to transform one or more other compounds, perhaps through only a few steps without being able to derive either energy or growth from the process. The phenomenon, which occurs widely in nature and is probably more significant in the degradation of xenobiotic compounds [6], embraces a number of slightly different mechanisms. Co-metabolism can result from a simultaneous attack on the growth substrate and the non-utilised substance by the same enzyme or sequence of enzymes. In this case the enzymes show a broad rather than a narrow specificity. In other instances, the co-metabolic attack may occur through the activity of enzymes not directly associated with the catabolism of the growth substrate.

Communities of Micro-Organisms

Most laboratory studies of metabolic pathways have been made with mono-cultures and mono-substrates; natural habitats including sewage treatment plants, however are much more complex containing many species of micro-organisms and many substrates. The microbial composition of activated sludge and percolating filter film has been studied [7, 8] and the dominant heterotrophs were found to be species of *Achromobacter, Alcaligenes, Arthrobacter, Bacillus, Comamonas, Flavobacterium, Micrococcus, Nocardia*, and *Pseudomonas*. Although the abilities of individual isolates to metabolise various compounds have been determined, very little is known about the consequences of growth and co-existence of mixed microbial populations in relation to biodegradative processes. It is probable [9] that beneficial relationships exist between different populations resulting in the formation of structural microbial communities which is some cases constitute units which are better equipped to exploit a given set of environmental conditions. Slater [9] identifies various sorts of microbial community with different potentials for degrading complex natural products and xenobiotic compounds.

The simplest community contains a species which synthesises and excretes a product required by a second species. In some cases the interaction is reciprocal, the second species providing an essential nutrient for the first. In other examples the second species removes a self-toxic and end-product formed by the first.

Another sort of community involves a combined metabolic attack on the substrate; only the combined activity of several species achieves complete degradation in a reasonable time. Although a compound (e.g. linear alkyl benzene sulphonate, or alkyl phenol ethoxylate) may be degraded at a moderate rate in nature, it is sometimes difficult to isolate individual species capable of effecting complete degradation or of achieving the rate in the natural environment.

Finally, other communities, in which complete mineralisation occurs, are based on co-metabolism. A number of possibilities exist, the most common being the co-metabolic formation, from the xenobiotic by one species, of an intermediate product which is completely oxidised for growth by a second species. The first species grows at the expense of a second substrate not available to the second species.

Adaptation to New Chemicals

Much interest centres on the behaviour of new chemicals in the environment; in particular, whether populations in rivers, waste-water treatment systems, etc. can adapt to degrade the novel compounds when discharged to the environment. A great deal of effort, both practical and theoretical, continues to be spent by microbiologists and biochemists on the various mechanisms by which micro-organisms are able to develop new catabolic activities and how rapidly such activities can be acquired. These studies, made chiefly with pure cultures, not only reveal interesting details about evolutionary changes but may also be useful in application to waste-water treatment systems for removing xenobiotic compounds more quickly and efficiently.

There are two general mechanisms by which acquisition of new degradative abilities can be achieved; first, the adaptation of existing catabolic enzymes, including those involved with transport and regulatory mechanisms and, second, the evolution of complete metabolic pathways. In the first case, broadly specific enzymes are altered by mutation so that refractory new compounds, usually structurally similar to degradable substrates, are also taken through the cell wall and degraded.

In the second case three distinct hypotheses have been proposed for the evolution of whole metabolic pathways; to some extent the two older established propositions are in opposition. The stepwise retrograde mechanism [10] involves gene duplication and subsequent modification of one copy to produce an altered protein which can catalyse the next step in the pathway. The proposal by Wu et al. [11] suggests that new pathways evolve from existing ones by alteration of enzymes with slight activity towards the metabolites of the new pathway. There is evidence, based on the nature and activity of enzymes in similar and disparate pathways, in support of both proposals.

The third mechanism, put forward by Slater and Somerville [12], is similar to that of Wu et al. but operates at the level of metabolic interaction between microbial communities. There is a much greater chance of there being an existing enzyme with fortuitous activity towards the new substrate in the larger genetic pool of many different species than in a single species. Similarly, the chances of locating a second enzyme to transform the product of the first reaction is high, but this enzyme is unlikely to be present in the same organism as the first. Degradation would then be the result of collective activity of a metabolically structured community. It is thought that genetic transfer, mediated by non-specific phage, plasmid or transformation, could eventually result in all the relevant genetic information of the new pathway existing within a single species.

Examples of mechanisms by which xenobiotic compounds are degraded, both before and after adaptation of populations, abound in the literature. Particularly informative is the Symposium on "Microbial Degradation of Xenobiotics and Recalcitrant Compounds" [13] (especially papers by Cain, Dagley, Fewson, Harder, Hutzinger, Knackmuss, and Williams) and a recent essay by Neilson [14].

Influence of Molecular Structure on Biodegradability

It is to be expected that xenobiotic compounds would stand a greater chance of being biodegraded the nearer their structure is to that of naturally-occurring compounds; reality, however, is rather complex. From the wide literature on the subject which includes a whole range of techniques, pure and mixed cultures, some general conclusions may be drawn [4, 15] as to the effect of molecular structure of a compound on the ease with which it is biodegraded. The reported data are such that the conclusions can only be tentative, and any general rules which can be established have many exceptions.

Few attempts at establishing quantitative relationships between structure (or some physico-chemical property) and biodegradability (QSBR) have been reported, and in those only relatively small groups of similar chemicals were used. These attempts have had varying success. For example, Yonezawa and Urushigawa [16] tried to relate the distribution coefficient between n-octanol and water (P_{ow}) of a number of aliphatic alcohols (C_4–C_8) with their biodegradability, expressed as a first order decay constant. For the five primary alcohols there was a linear relationship with the C_8 alcohol degrading faster than the C_4, but for the other series of alcohols the relationships were not so simple. On the other hand, Paris et al. [17] obtained good correlationship between the second order constant of the alkaline hydrolysis of ten esters and their second order biodegradability constants (see below).

Even fewer attempts to obtain more general QSBRs have been reported and these have not been very successful. The EPA [18] used a very wide definition of biodegradability, which was expressed simply as "yes/no," and tried to relate this to structure, as expressed by a modified Wiswesser line-formula notation. Of the 430 compounds used in the final comparison, over 92% of degradable compounds were accurately predicted ($P > 0.7$) from the discriminant equation, but only 68% of non-biodegradable compounds were predicted as such. The authors offer satisfactory explanations of the bias towards biodegradation (insufficient non-degradable compounds were included) and reasonably suggest that improvements could be made by involving partition coefficients and parameters such as molecular connectivity.

The only other attempt found in the literature was that by Mudder [19], who related the kinetic removal rate (mg COD/g biomass h) of 54 mono- and disubstituted benzenes to their structure as expressed by different combinations of 14 structural descriptors and six physico-chemical variables. (The kinetic data were obtained by Pitter [20] using activated sludge acclimatised to the test substance for 20 d in a fill-and-draw system having a sludge age of 5 d.) Comparisons between observed kinetic rates and those obtained from the discriminant equations for compounds both within and outside the data-set were not particularly good. Mudder suggests explanations for some of the discrepancies on the basis of exceptional ease of initial hydroxylation of some compounds and of inhibition of the competent organisms by other compounds, factors not accounted for in the structural descriptors.

It would seem that these approaches should be continued, but that much needs to be done before reasonably predictive equations are attained. To take the simple example of D-glucose, it has been found [21] that unacclimatised sludges taken from a variety of sources degrade the sugar at rates varying from 20 to 160 mg/g dry sludge h, while acclimatised sludges degraded glucose at 180–1000 mg/g h.

Kinetics of Degradation

The knowledge that a compound can be degraded is insufficient by itself for hazard assessment; an indication of its rate of removal in the various compartments of the aquatic environment is also required. Pentachlorophenol can be degraded only by rare micro-organisms and mixed cultures under special conditions [22] but would not be put in the same category as, say, 2-chlorophenol which is fairly rapidly metabolised by normal, unacclimatised sewage micro-organisms in standard die-away tests.

Substrates are removed from solution by micro-organisms at rates which can be predicted by equations based on Michaelis-Monod kinetics. The usual form of the equation is

$$-\frac{d[S]}{dt} = \frac{\mu_m[B][S]}{Y(K_s+[S])} \tag{1}$$

where [S] = substrate concentration (mg/1)
    [B] = biomass concentration (No. or mg/1)
    μ_m = maximum specific growth rate (h^{-1})
    K_s = substrate concentration at one half μ_m (mg/1)
    (= saturation constant)
    Y = yield coefficient = cells or mass produced per unit of substrate metabolised.

The equation has successfully been applied to waste-water treatment processes for single and multi-substrate [e.g. 23] and to much more dilute systems – rivers, lakes,

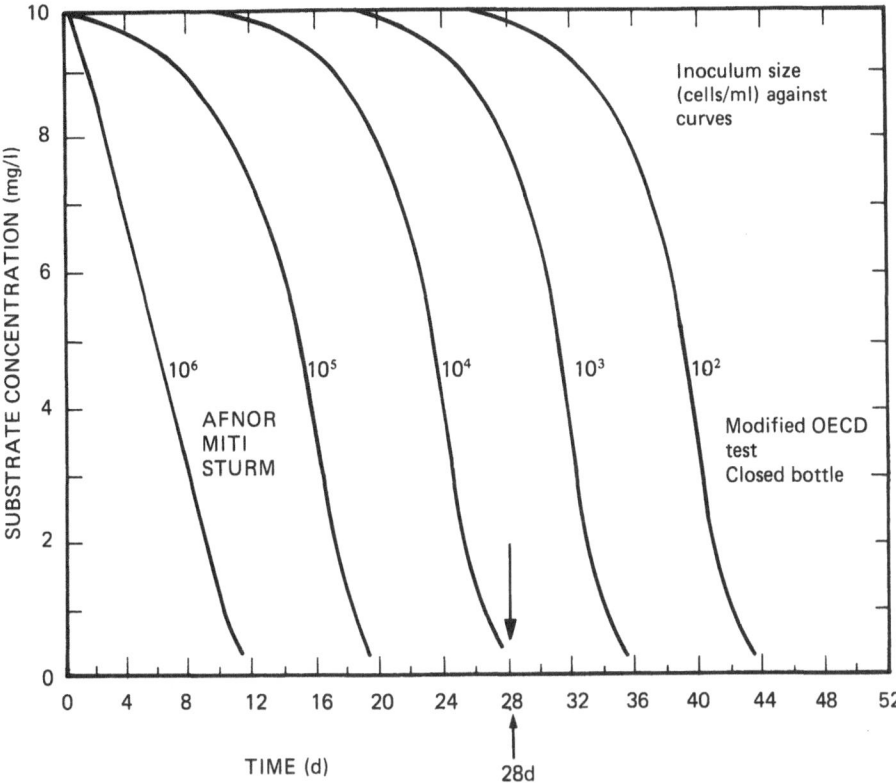

Fig. 1. Theoretical die-away curves: μ_{max} = 0.35 d^{-1} (by permission of Academic Press Ltd. Ref. 26)

seas [e.g. 24, 25]. In the latter case, Paris [25] has used a simplified equation

$$-\frac{d[S]}{dt} = \frac{\mu_m[B][S]}{YK_s}$$ (2)

since $[S] \ll K_s$ in river water, and has likened this to the "second-order" equations used to describe the kinetics of chemical reactions. The term μ_m/YK_s is termed a second-order biotransformation rate constant.

Currently results of biodegradability tests are usually reported as percentage removal of the organic substance originally added. This is not very fruitful for predictive purposes when applying results of laboratory tests to the environment. Evaluation from laboratory tests of the second-order rate constant, or, better still, of μ_m and K_s associated with xenobiotic compounds will be a considerable aid in making the necessary predictions. A kinetic interpretation [26] of the results of OECD/EEC die-away tests has demonstrated the effect of biomass concentration on the pattern of removal of a test substance (Fig. 1) and has helped in making predictions as to whether a substance will be removed in waste water treatment. A promising technique called "repetitive die-away" [27] is designed to reduce the effect of [B] in equation (1) and to increase the concentration of competent organisms, allowing more accurate assessments of μ_m to be made.

Using a generalised form of the logistics function, Larson [28] has recently shown that the results of screening tests usually underestimate the potential for degradation in the environment; in the extreme the limitations of the test system can lead to no removal of a compound shown subsequently to be adequately removed in the environment. Further work on this topic should be very fruitful.

Test Methods

General

As the activity of the chemical manufacturing industry increased over the past few decades so also did the interest in assessing the biodegradability of the large number of xenobiotic chemicals produced. There are only a few basic principles by which biodegradability may be assessed but, because of the wide ranging interests of individual organisations and workers in the field, many variations on these themes have been reported. Some of these methods have been summarised by the EPA [29], Gilbert and Watson [30] and de Kreuk and Hansveit [31, 32].

As indicated earlier, the methods are based on aspects of microbial growth and metabolism, namely, determination of
(a) removal of the test substance *per se* or as a summary parameter such as dissolved organic carbon (DOC) or chemical oxygen demand (COD);
(b) uptake of oxygen;
(c) evolution of carbon dioxide;
(d) increase in biomass.

Direct Methods

Specific analyses for a chemical in test media over a set period will establish whether the chemical has changed in some way during the test. It will show, conclusively, only the occurrence of primary biodegradation. The classic example of this is the application of the methylene blue colorimetric determination to the degradation of anionic surfactants. Shortening of the alkyl side chain, by ω-oxidation followed by β-oxidation, to form carboxylic acids, yields products in the reaction with methylene blue which are not extracted by chloroform (an essential step in the analytical procedure). Similarly, desulphonation, the alternative initial step in the degradation, yields an intermediate incapable of reacting with methylene blue. It was shown by Bock and Stache [33] that over the same incubation period, less DOC (and ^{14}C) than MBAS was removed, indicating that lowering of methylene blue reactivity demonstrates that only primary degradation has occurred. It so happens that the removal of MBAS also lessens the foaming properties and toxicity to aquatic organisms: this has been termed environmentally acceptable biodegradation [30]. A similar situation holds with the alcohol ethoxylate and alkylphenol ethoxylate non-ionic surfactants [33].

Apart from the fact that it is important to know whether a chemical can be completely mineralised (ultimate biodegradation) rather than only undergoing a change(s) in its molecular structure, there is the practical question of cost in establishing and applying specific analytical methods for all chemicals to be tested. It is thus now more usual to analyse for a summary parameter such as *DOC* or *COD*. Under these circumstances the test substance must be present as the sole source of organic carbon. Bacterial cells and other suspended matter added with the inoculum must, of course, be removed before analysis by centrifugation or by filtration through a suitable membrane (pore size 0.2 μm) which does not remove from, or add to, the filtrate organic carbon. The minimum concentration of test substance which can be used is limited by the sensitivity and precision of carbon analysers; with analysers currently available initial concentrations of DOC should be at least 5 mg/l. The maximum concentration of test substance is limited by its inhibitory properties towards bacteria, though in legislative test methods an upper limit of 40 mg C/l, or in some cases 400 mg C/l, is stipulated. The concentration of DOC remaining at the end of incubation, corrected for that in a control flask containing no test substance (to allow for carry-over by and changes in the inoculum), is then a measure of unchanged test substance plus any organic intermediates.

Theoretically 100% of the initial DOC can be removed. Whether all moieties of the molecular structure of a substance have been attacked can usually be decided from a consideration of the structure and from the proportion of DOC removed. In some cases it may be necessary to examine culture filtrates for possible intermediates. The OECD Chemical Group [34] decided that removals of test substance (specific analysis) of >80% and of DOC of >70% generally indicated that all moieties of the molecules were undergoing attack.

Indirect Methods

The uptake of oxygen resulting from the oxidation of a test substance in the presence of a bacterial inoculum is the next most commonly used method; it has the advantage, along with the carbon dioxide evolution method, of being applicable to the study of insoluble substance, whereas the DOC die-away method is not. (Methods for substances insoluble in water have not yet been subjected to the same scrutiny and calibration-testing as have soluble substances. This is partly because the investigation of the fate of insoluble substances was originally considered of lower priority, since they were thought to have less environmental effect, and partly because of difficulties of ensuring uniform dispersion in the culture media.) Enclosed, two-phase manometric respirometers are employed, offering the opportunity of assessing the biodegradability of volatile chemicals, provided that certain safeguards are taken.

As with the DOC die-away method, control reaction vessels containing inoculum but no test substance must be included to take account of endogenous respiration. In both indirect methods it is advisable to choose inocula which, though active, have relatively low endogenous respiration rates, since both methods depend on differences between two measured values. Investigators therefore have tended to use higher concentrations of test substance (around 100 mg/l) when measuring oxygen uptake, as compared with the DOC die-away method. Unaccountably, only 10–20 mg/l of test substance is used in the carbon dioxide-evolution method.

Assessment of biodegradability is made by expressing the oxygen uptake and carbon dioxide evolved as percentages of the respective theoretical values, calculated from stoichiometric equations or, if impure, from the measured COD (for oxygen uptake) or measured carbon content (for carbon dioxide production). A disadvantage of this is that the percentage values are unlikely to reach 100% in the period of the test (they are often much less than this in practice) because some of the substrate carbon is synthesised into new cells and is not converted to mineral products. The proportion of substrate carbon divided between cell synthesis and respiration differs considerably between species and between states of growth. The amount of organic carbon removed from solution can, however, often be determined at the end of the test as a confirmatory procedure. To allow for cell synthesis, the OECD Chemical Group [34] recommended that complete degradation (ultimate) can be assumed if >60% of the theoretical value(s) is recorded. Also, it is normal practice to include in each batch of tests, of any type, a standard chemical so that the behaviour of the inoculum with respect to cell synthesis may to some degree, be monitored.

An important factor in the oxygen-uptake method is that appropriate corrections should be made to take account of any oxidised nitrogen formed during the test period by both the control and flasks containing test substances. Nitrogen-containing substances may be oxidised to nitrate rather than being hydrolysed to ammonia and some chemicals may inhibit nitrification (not inhibited in the control).

Growth Methods

Growth, measured as turbidity, has been used to assess biodegradability of test substances present in media as the sole source of carbon. This is commonly used in conjunction with enrichment techniques [35], starting with a readily degradable substrate present in addition to the test chemical. When the culture becomes turbid, a small portion is sub-cultured to another medium containing a lower proportion of the alternative degradable substrate and more of the test chemical. This procedure is continued until the final medium contains only the test chemical as source of carbon. Visible growth in the final medium of the series shows that the test substance supports growth and was therefore biodegraded.

Alternatively, growth is measured as increase in protein concentration of the culture. In one version [36], the test chemical is added as the sole source of N, P or S; the carbon is provided by a number of other degradable organic substances.

The media used must be completely free from N, P or S, whichever element is to be provided by the test substance. In order for growth to occur, the micro-organisms must obtain the element by degrading the test substance.

Application of Radiolabelled Substrates

[14]C-labelled substrates have frequently been used to help in the elucidation of many metabolic pathways and are increasingly being used in biodegradability

testing. The positioning of the labelled atom(s) in the molecule is important and can be used selectively to distinguish which moiety of the molecule has been attacked. For example, in aromatic compounds with side chains $^{14}CO_2$ evolved from ring-labelled preparations would indicate ring opening, while $^{14}CO_2$ from labelled side chains shows only aliphatic oxidation. In both these cases primary biodegradability is indicated; uniform labelling would have to be used to determine ultimate biodegradability.

In addition to trapping and measuring $^{14}CO_2$ evolved from an inoculated test medium, further information can be gleaned by filtration of the medium to remove bacterial cells and subsequent counting of the solids on the membrane and in the filtrate. (Two practical details are important: the filtrate should be acidified and aerated to collect any inorganic carbon at the end of incubation and any adsorption of ^{14}C on the membrane, etc. must be accounted for.) Thus the disappearance of substrate ^{14}C can be followed to complement the $^{14}CO_2$-evolution measurements.

The cost of radio-labelling a wide variety of chemicals specifically for the purpose of testing their biodegradability makes the technique applicable only under special circumstances, but it does allow rates of biodegradation to be determined under conditions ($\mu g/l$) more nearly simulating those in the environment.

Composition of Media

Provided that essential mineral macro-nutrients, such as N, P, S, K, Na, Fe, Ca, Mg, are present in adequate concentration and correct proportions in addition to an assimilable carbon source, most species of bacteria can grow in a variety of media. Indeed, successful biodegradability tests have been reported in which many varied media have been used. It has been found [37] that the addition of a number of trace elements results in increased degradation of some chemicals, so that trace elements are often added to media, although in many cases sufficient of these micro-nutrients are added with the inoculum.

Another requirement for complete growth (and complete utilisation of test substance) is that the pH value of the media stays within the range 6–8 throughout the test. It seems probable that some reported cases of partial degradation may have been due to development of a hostile acid or alkaline condition. Thus, the medium must be sufficiently buffered to cope with any production of acid or base from the test substance or of acid from the oxidation of ammonia.

A minority of bacterial species has specialised needs for co-factors and it is prudent to include eight members of the B-group of vitamins, as such or as yeast extract, to satisfy this requirement. When cometabolic degradation is to be assessed, the medium must include degradable organic compound(s).

Finally, water used in preparing media must be free from toxic metals and contain concentrations of organic carbon as low as possible, preferably < 1 mg C/l.

Inocula

The nature and quantity of the inoculum are very important in biodegradability assessments and the inoculum is probably the biggest single factor in the success of a batch test. One approach is to inoculate the medium with one of a number of versatile pure stock cultures [38] for either growth or die-away tests. Likely species to try are those of *Aerobacter, Escherichia, Pseudomonas, Nocardia,* and the fungus *Aspergillus*. However, it is much more usual and rewarding to take an inoculum from a source which contains a wide variety of species, since mixed microbial populations have been found to attack a wider range of chemicals than does any single species. Further, using mixed populations is more relevant to the environment.

River water has often been used as a source of micro-organisms, especially if knowledge was required of what would happen if a test substance were discharged to a river environment. The water can be amended with extra necessary nutrients or the organisms in the river water can be concentrated as, for example, by the ultra-filtration procedure and subsequent re-suspension, as described in the AFNOR method [39].

Probably the most common sources of inocula are good-quality *effluents* or *activated sludge* from sewage treatment works. The mixed microbial population naturally occurring in domestic sewage – from the gut flora and free-living aquatic bacteria – is encouraged and developed in sewage treatment processes specifically for the biodegradation of waste waters. Industrial waste waters are often treated in admixture with domestic sewage, so that the bacteria are also exposed to a wide range of synthetic chemicals. However, care should be taken when inocula from treatment plants receiving a significant amount of one particular industrial waste are used, because these bacteria may have become adapted to the chemicals present in that waste. If the substances to be tested are similar in structure to those present in the industrial waste, the adapted sludge or effluent may have an unusually high activity towards the test substance. In such circumstances inocula from another source should be obtained. Activated sludges, of course, contain organisms other than bacteria; protozoa, for example, are important in the treatment processes since they are responsible for producing clear effluents by scavenging non-flocculent bacterial cells, but their role in biodegradation is probably only minor, if any.

Soil contains many types of bacteria, aerobic, anaerobic, and facultatively anaerobic, and aqueous extracts of soil have often been used as inocula. Care must be taken to choose well balanced soils and to avoid those with extreme contents of clay, sand etc. Also, soils recently fertilised or treated with pesticides should not be used since, in the latter case, adaptation of bacteria could have taken place and might affect their behaviour to the test substance.

Some workers have employed inocula from a *mixture of sources,* e.g. a mixture of river water, soil extract and sewage effluent. In the Japanese MITI respiration methods [40] the source of inoculum is more complex. Samples of sludge, mud, river, and sea waters are collected from at least ten sites within a country in areas where a variety of chemicals is known to be used and discharged to the environment. The supernatant of the settled mixture from all the sites is aerated and treated in the fill-and-draw activated sludge mode, discarding 1/3 of culture volume daily and feeding with an equal volume of 0.1% each of glucose, peptone, and phosphate (at pH 7.2). After one month the "activated sludge" is ready for use as an inoculum but is discarded after a further 3 months.

Turning to the *concentration of the bacteria,* added via the inoculum, if a test substance is degraded by many of the species present in the inoculum even low initial numbers of the competent organisms will soon multiply in the presence of the potential substrate and oxidation will be rapid. If, however, there are only a few species, present in low numbers which can metabolise the test substance, a lag will occur during which the competent species will multiply until a sufficient number are present to make a significant reduction in the concentration of the test chemical. (During the lag period enzyme induction and other adaptive processes may also occur.) In this situation biodegradation depends much more upon the duration of the test than in the former case. Painter and King [26] have defined the minimum specific growth rate required by competent species of bacteria for various initial bacterial concentrations in order to obtain positive results in tests of defined duration. Making the improbable, but "fail-safe" assumption that all organisms in the inoculum were competent, it was shown that, even with the highest concentration of bacteria (see Table 2 later) used in recommended OECD tests, no substance would degrade under the test conditions which would not be found subsequently to degrade in activated sludge treatment plant operated with sludge ages of 4–6 d (equivalent to specific growth rates of 0.25–0.167 d^{-1}).

Blok and Booy [27] have advocated the use of a batch repetitive die-away test, in which up to three weekly additions of the test substance are made. By this means the numbers of competent organisms are increased and each successive rate of biodegradation becomes less dependent on the numbers of organisms in the initial inoculum.

Acclimatised Inocula

Most of the foregoing relates to batch tests for *ready biodegradability,* which by definition (see Glossary) must be assessed using inocula not previously adapted to the test substance. If, however, batch tests are to be used to assess *inherent biodegradability,* then any form of pre-adaptation or acclimatisation may be used. The original Sturm [41] test included a preacclimatisation of sewage to the test substance of 15 d, while Pitter [22] exposed activated sludge in a fill-and-draw system to increasing concentrations of the test substance over a period of 20 d, before determining the rate of degradation of the test substance by some of the acclimatised sludge. Zahn and Wellens [42] reported greatly increased rates of degradation of nine chemicals by repeating the Zahn-Wellens test [43] using sludge taken from the first attempt. Gerike and Fischer [44] have adapted sludges by exposing them to test chemicals in both the Zahn-Wellens [43] and OECD confirmatory methods. It is considered that the 42 d period allowed in the AFNOR die-away method [39] is sufficient to permit acclimatisation.

No evidence seems to have been reported as to why the various periods of acclimatisation have been adopted and, as yet, no agreed ways of acclimatisation exist. Little is known about the effects of factors, such as time and pattern of exposure, concentration of the test substance, presence of other substrates, on the adaptive processes.

Because some test substances are degraded abiotically, it may be necessary to set up test flasks to which all ingredients are added except an inoculum. In extreme cases, it may be desirable to sterilise the flask and medium to ensure that no microbial action occurs.

Duration and Extent of Degradation

When studying a chemical from the academic point of view, any attempt to assess biodegradability in a batch system can, of course, be continued as long as the experimenter wishes, though it would seem to be pointless to continue for too long. However, in the interest of standardisation and for legislative purposes, limitations have to be imposed on the duration of such attempts. Before the OECD Chemicals Group issued its report [34], the prevailing methods stipulated various periods. For example, the BOD method occupied 5 d, while Fischer's "Closed Bottle" version [45] ran for 30 d and the MITI respirometric method permitted only 14 d. The OECD die-away method for detergent degradation [46], enshrined in EEC Directives [47–49] allows 19 d, though the Sturm method [41] continued for 27 d and the AFNOR method [39] as long as 42 d.

In considering this aspect, the OECD Chemicals Group [34] decided to recommend that the length of time for batch biodegradability assessments should be 28 d, but that if degradation had started by the 28th day it was only sensible to continue for a further short period to ascertain whether a plateau of removal was attained.

Along with this recommendation on duration was another on the extent of removal of test substance, as such or as DOC, which indicated full degradation. Based on collective experience, it was decided that complete primary degradation was deemed to have occurred if 80% test substance had been removed and that for ultimate degradation 70% DOC had been removed. The corresponding value for the oxygen uptake and carbon dioxide indices was 60%. Further, a restriction was introduced to allow for adaptation or acclimatisation occurring within the test period. Thus, biodegradation was deemed to have begun when 10% removal was detected and the "pass" values must be attained within 10 days of this occurring.

A theoretical approach [26], based on the Monod equation of bacterial growth, and taking into account the different initial concentrations of "total" bacteria in the various test methods, has indicated that the OECD recommendations are "fail-safe" as regards assessment of hazards to the environment.

Standard Tests

Working with variations of the parameters described above, different laboratories developed their own methods to suit their particular purposes, often choosing media which were in general use or easily available in their laboratory. The inoculum also tended to be taken from readily accessible sources in the vicinity of the laboratory or prepared in some idiosyncratic way. Thus, there was a proliferation of methods having similar purposes and often giving similar results. For the most part there were only trivial differences between the various methods which were probably not significant to the final outcome. However, it was generally agreed that some degree of uniformity in methodology was necessary in view of the large

Table 1. Types of biodegradability tests

Type	Method	References
Screening	Closed bottle	[45, 50]
	Modified OECD[a]	[50]
	MITI I (Japanese Ministry of International Trade & Industry)	[40, 50]
	EEC respirometric	To be published
	Sturm	[41, 50]
	AFNOR (Association Francaise de Normalisation)	[39, 50]
	ISO (International Organisation for Standardisation)	[51]
Inherent	Zahn-Wellens	[43, 50]
	SCAS (Semi-continuous activated sludge)	[35, 50, 58]
	MITI II	[40, 50]
Simulation	Activated sludge: Husmann, Porous Pot	[35, 50, 61]
	Biological filters: rotating tube	[35, 60]
	River: Karlsruher	[59]

[a] Modified from the OECD detergent method
In other cases in the text, the word "modified" precedes the name of the test to denote that the test has been altered from the original

number of new chemicals which should be examined before coming on to the market, since such chemicals might cause environmental hazards.

In considering the available methods and results of calibration exercises of some of them, the OECD Chemicals Group [34] recognised three types of test. Each type of test is linked to a type of biodegradability, as indicated in Table 1.

The purpose of the simplest test methods, known as *screening tests,* is to obtain an initial indication of the biodegradability of a chemical quickly and with a minimum of time and effort. By definition, they identify substances which are *readily biodegradable* and are designed to be "fail safe." Positive results will identify substances which will certainly be easily degraded under natural conditions in the aquatic environment, where there will be a wider range of micro-organisms and other complex factors inter-playing and combining to remove the chemical. Thus, if a substance is degraded in a screening test no further testing for degradation is normally required. However, those substances which fail are not necessarily non-biodegradable but must be examined further.

Screening Tests

Screening tests listed in Table 2, with media listed in Table 3, are batch tests usually having the test substance as the only source of carbon and energy. It is dissolved in a mineral salts medium and the solution is inoculated with a relatively small number of unacclimatised micro-organisms. Test vessels are incubated in the dark at 20°–25 °C for 28 d. Samples are taken for analysis, or measurements made, at sufficiently frequent intervals to allow the calculation of the degree of removal (or of oxygen uptake or carbon dioxide produced) in the 10 d period ("window") after degradation starts. Suitable blank controls are run with each test to allow for the activity and carbon content of the inoculum and substances

Table 2. Details of biodegradability tests

Method	Inoculum (volume or weight per litre of medium)	Concentration of bacteria (organisms/ml)	Parameter determined	Reference
Ready				
Closed bottle	1 drop effluent	0.25×10^2	Oxygen uptake	[45]; OECD [50]
Modified OECD	0.5 ml effluent	$0.5 - 2.5 \times 10^2$	DOC removed	OECD [50]
BOD (USA)	1–2 ml effluent	$0.1 - 1 \times 10^3$	Oxygen uptake	
BOD (UK)	5 ml effluent	$0.5 - 2.5 \times 10^3$	Oxygen uptake	
Sturm	1% supernatant of homogenised activated sludge	$10^4 - 2 \times 10^5$	CO_2 evolved (DOC removed)	[41]; OECD [50]
AFNOR	Resuspended bacteria filtered from surface water/effluent and counted	$5 \pm 3 \times 10^5$	DOC removed	[39]; OECD [50]
MITI I	30 mg suspended solids (ss) of activated sludge grown on glucose-peptone medium	$2-10 \times 10^5$	Oxygen uptake (DOC removed)	[40]; OECD [50]
EEC respirometric	30 mg ss of activated sludge	$2-10 \times 10^5$	Oxygen uptake (DOC removed)	To be published
ISO	0.5 ml effluent – 30 mg ss of activated sludge	$0.5 - 2.5 \times 10^2$ to $2 - 10 \times 10^5$	DOC removed	[51]
Inherent				
Bunch-Chambers	100 ml settled domestic sewage	$10^5 - 10^6$	Specific analysis	[35]; [54]
MITI II	100 mg ss of activated sludge	$0.7 - 3.3 \times 10^6$	Oxygen uptake	[40]; OECD [50]
Zahn-Wellens	1 000 mg ss of activated sludge	$0.6 - 3 \times 10^7$	DOC removed	[43]; OECD [50]

of known biodegradability such as aniline, sodium benzoate, sodium acetate, glucose, are also tested to check that the inoculum is active. In some methods extra vessels can be set up containing both the test and standard substances in order to ascertain whether inhibition of the inoculum has occurred. Should lack of degradation of test substance be thought to be due to inhibition, the test may be repeated with a lower concentration of test substance. The *Closed Bottle* test [50] is prepared in the same manner as the traditional 5-day BOD test, using air-saturated dilution water. The test substance is added at ~ 2 mg/l and the inoculum is only one drop of a filtered, good-quality sewage effluent or soil extract added to 1 l of medium. Blank controls without inoculum are prepared to check that the dilution water used has little bacterial activity, and with inoculum to monitor its activity.

The other oxygen uptake method, the *modified MITI I* method [50], permits the continuous recording of the oxygen consumption by means of an automatic manometric respirometer. The medium contains the same constituents as the BOD dilution water but at threefold concentration and the specially prepared inoculum is added at a very much higher concentration. Even with this higher buf-

Table 3. Composition of media. (Values expressed as mg/l, except total PO_4, which is $\times 10^{-3} M$)

Constituent	Closed bottle	Modified OECD	Respirometric		Sturm	AFNOR	ISO
			MITI	EEC			
KH_2PO_4	8.5	8.5	25.5	85	17	300	85
K_2HPO_4	21.8	21.8	65.3	217.5	43.5	–	217.5
$Na_2HPO_4 \cdot 2H_2O$	33.4	33.4	133.8 (12)	334	66.8 (7)	2,000	334
Total PO_4	0.375	0.375	0.94	3.75	0.59	7.8	3.75
$MgSO_4 \cdot 7H_2O$	22.5	22.5	67.5	22.5	22.5	50	22.5
$CaCl_2$	27.5	27.5	82.5	27.5	27.5	37.8	27.5
NH_4Cl	1.7	20	5.1	25	3.4	–	25
$(NH_4)_2SO_4$	–	–	–	–	40	300	–
NH_4NO_3	–	–	–	–	–	150	–
$FeCl_3 \cdot 6H_2O$	0.25	0.25	0.75	0.25	1	–	0.25
$FeCl_3 \cdot EDTA$	–	0.1	–	0.1	–	–	0.1
$MnSO_4 \cdot 4H_2O$	–	0.04	–	0.04	–	–	0.04
$(NH_4)_6Mo_7O_{24}$	–	0.035	–	0.035	–	–	0.035
H_3BO_3	–	0.057	–	0.057	–	–	0.057
$ZnSO_4 \cdot 7H_2O$	–	0.043	–	0.043	–	–	0.043
Yeast extract	–	0.15[a]	–	0.15[a]	–	5	0.15[a]
Trace elements	–	–	–	–	–	Yes[b]	–

[a] Or 1 ml/l of a solution containing, per 100 ml: biotin, 0.2 mg; nicotinic acid, 2.0 mg; thiamine, 1.0 mg; p-aminobenzoic acid, 1.0 mg; pantothenic acid, 1.0 mg; pyridoxamine, 5.0 mg; cyano-cobalamine, 2.0 mg; folic acid, 5.0 mg

[b] 1 ml/l of a solution containing, per 100 ml: $FeSO_4 \cdot 7H_2O$, 0.1 g; $MnSO_4 \cdot H_2O$, 0,1 g; K_2MoO_4, 0.025 g; $Na_2B_4O_7 \cdot 10H_2O$, 0.025 g; $Co(NO_3)_2 \cdot 6H_2O$, 0.025 g; $ZnCl_2$, 0.025 g; NH_4VO_3, 0.010 g

Numbers in brackets are the number of molecules of water of crystallisation

fering capacity, it is wise to check pH values before and after incubation. (Another version of this method, currently being subject to a calibration exercise by the EEC/OECD, has a tenfold higher buffering capacity than the BOD dilution water (Table 3) and normal activated sludge as the inoculum.) In order to obtain greater precision, it is usual to add as much as 100 mg test substance per litre, which sometimes leads to inhibition. Aniline is the standard for checking the activity of the medium.

In both these oxygen uptake methods, it must be ascertained whether the ammonium salts in the medium and any nitrogen in the test substance are oxidised to nitrite and/or nitrate during incubation. Any such oxidation must be taken into account in calculating the oxygen consumed by the test substance and in expressing it as a proportion of the corresponding theoretical consumption.

The *modified OECD die-away* method [50] (derived from that used in detergent tests) employs an inoculum 0.5 ml of a good quality effluent per litre medium, which is similar to BOD dilution water but is fortified with trace elements and vitamins. Test vessels are covered, e.g. with aluminium foil, and shaken to maintain the concentration of dissolved oxygen. Samples are filtered through membranes to remove bacteria and other suspended solids prior to analysis for DOC in the filtrate. Care must be taken to ensure that the filtration step does not add extraneous organic carbon from the membrane, nor remove DOC from the sample.

The *modified AFNOR* method [50] is similar to the modified OECD method in that DOC is determined on filtered samples, but the medium is somewhat different (Table 3). It has a greater buffering capacity, it contains two extra trace elements – cobalt and vanadium – and has a higher concentration of yeast extract (5 mg/l cf 0.15 mg/l). The inoculum used is quite different. A mixture of three sources of polluted surface waters and good quality sewage effluents, each containing at least 10^5 bacteria/ml, is ultra-filtered. The collected solids are re-suspended in a solution isotonic with the sources and the turbidity is measured. The concentration of cells is read from a calibration graph prepared from standard suspensions of *Pseudomonas fluorescens* and is adjusted to $5 \pm 3 \times 10^7$ cells/ml. This suspension is diluted 1 in 100 in the final medium.

The method adopted by the *International Organisation of Standardisation* [51] is a combination of the two preceding methods and permits inocula ranging from 0.5 ml effluent/l to 30 mg/l of activated sludge solids equivalent to 10^5–10^6 cells/ml, as in the EEC version of the MITI method.

The *modified Sturm* method [50] is carried out by slowly bubbling CO_2-free air through an inorganic medium containing the test substance at 10–20 mg C/l. The medium, based on the BOD dilution water, contains extra ammonium salts, has twice the buffering capacity but contains no added trace elements or vitamins. Sufficient of both of these is probably added via the inoculum, which is settled supernatant obtained by homogenising activated sludge mixed liquor, previously aerated for 4 h without feed. The supernatant normally contains 10^6–2×10^7 "colony-forming units"/ml and is added to the medium at the rate of 1 ml per 100 ml. Carbon dioxide produced in the test vessels is precipitated as barium carbonate in a series of three baryta traps and the amount produced is determined by titrating the remaining barium hydroxide every 2–5 d in the trap first in line; the traps are moved up and another is added.

Because of the varied concentration (25–10^6/ml) of bacteria present in the screening tests, more substances are "passed" by the AFNOR, ISO, and Sturm methods than by the Closed Bottle and modified OECD methods [52, 53]. The MITI I method holds an intermediate position since, although the bacterial concentration is high, inhibition probably occurs more frequently by virtue of the high concentration of test substance.

There are some elaborations on these basic screening tests which make them more likely to yield positive results for those substances which are degraded only by cometabolic processes. For example, in the *Bunch-Chambers method* [54] (noted by the OECD Chemical Group) yeast extract is added (55 mg/l) as an alternative substrate to the test substance with 10% sewage as inoculum. At weekly intervals, 10 ml of the culture is trans-inoculated into fresh yeast extract-test substance medium. At each transfer the concentration of test substance is determined by specific analysis to assess the degree of primary biodegradation.

Another promising method is *Blok's repetitive die-away method* [55], which though not yet categorised probably falls within the definition of screening tests. It combines assessment of oxygen uptake and DOC removal and utilises the known volume of air space in a partially filled BOD bottle as a reservoir of oxygen, thus overcoming the limitation of the BOD bottle method. The stirred liquid phase in continuously re-aerated and the dissolved oxygen is in equilibrium with the oxygen in the gaseous phase. The removal of DOC and the oxygen uptake are determined at weekly intervals, using an oxygen electrode for the latter. If removal has occurred, more test substance is added and oxygen is replenished by opening the bottle and stirring. (Other details of this method are given under "Inocula".)

Other tests which could be classified as screening methods, e.g. growth tests, are not formally included in the OECD scheme since they have not been subjected to sufficient scrutiny.

Tests for Inherent Biodegradability

Other tests are required for the examination of substances which are not readily biodegradable: three such tests are included in the OECD scheme [50]. Originally the *Zahn-Wellens* [43] method was put forward as a screening method but, since experience showed that it was considerably less stringent (that is, "passed" more chemicals) than the five screening tests, it was re-classified as a test for inherent biodegradability. The method has been recently revised [56] and combined with the EMPA [1] method [57]. A much larger inoculum (200–1,000 mg sludge solids/l) and a higher concentration of test substance (50–400 mg C/l) are used than in screening tests. To allow for any adsorption of test substance onto sludge, a sample is taken 3 h after mixing, and further samples are taken at 1–3 d intervals for analysis of DOC, after filtration. There is a greater chance of nitrification occurring, because of the higher concentration of organisms, thus lowering the pH of the medium, and this should be checked daily and adjusted, if necessary. The higher concentration of the inoculum gives a greater opportunity of more test substances being degraded within the 28 d period of incubation.

The other recognised method for assessing inherent biodegradability is the *Semi-Continuous Activated Sludge* (SCAS) method [35], originally used by the SDA [58]. Domestic sewage and the test substance (at 5–20 mg C/l) are added daily to activated sludge in a suitable vessel (Fig. 2), which is aerated for 23 h. Sludge is then allowed to settle for ½–1 h, the supernatant is discarded or collected for analysis, fresh sewage plus test substance are added and the cycle is repeated. Control units receiving no test substance are operated simultaneously. Tests are usually continued for 3 months, or more, if necessary, during which time the concentration of DOC in the supernatants of test and control units are compared to ascertain whether, and how much, removal has occurred. It is assumed that the sludge in the test unit would oxidise the sewage components to the same extent as that in the control, so that any residual DOC, after subtraction of the control effluent value from the test effluent value, would be derived from the test substance. The pattern of removal over several weeks can usually help in distinguishing between adsorption and biodegradability, but any doubt can be dispelled by using the acclimatised sludge as an inoculum in a screening test, such as the Sturm or respirometer methods.

---

1 Swiss Federal Laboratories for Materials Testing and Research, St. Gallen

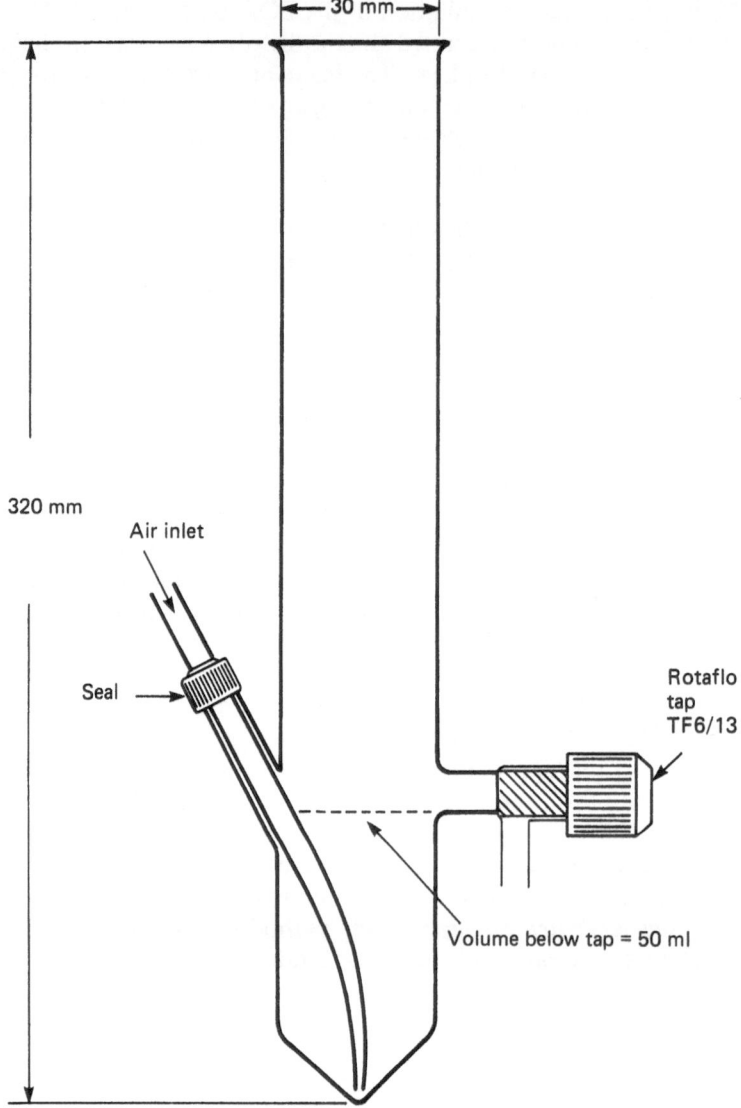

Fig. 2. Apparatus for semi-continuous sludge test (by permission of HMSO, London: Ref. 35)

The conditions in the SCAS method provide a good opportunity for biodegradation to take place. The mixture is inoculated freshly everyday with a wide variety of species by means of the addition of sewage, and the chances of cometabolism occurring are enhanced. During the 23 h aeration period the organic compounds in sewage are rapidly removed, leaving the test substance as the main potential substrate to be attacked, thus favouring selection and adaptation of bacteria capable of utilising the test substance. Lastly, because no sludge is deliberately wasted, slow growing species, not usually maintained in normal activated sludge (with sludge ages of about 6 d), will be retained in the SCAS vessel, further increasing the chance of degradation of the test substance.

The *Japanese MITI II respirometric method* [50] is less stringent than MITI I, since it employs a higher concentration of inoculum (100 mg/l) and 30 mg test substance/l, but it is much more stringent than the Zahn-Wellens and SCAS methods.

Simulation Test Methods

Substances which are degraded by the Zahn-Wellens or SCAS methods do not necessarily degrade readily in the environment: those substances which do not degrade in the inherent tests are considered to be non-biodegradable. To decide the fate of inherently degradable substances it is necessary to subject them to a test which simulates the environment to which they will be discharged. Tests simulating conditions in rivers e.g. the Karlsruher test [59] have been devised but little seems to have been reported on their use. (Soils are outside the scope of this chapter.)

Because the bulk of chemicals is discharged to sewers much attention has been given to simulations of activated sludge systems and to a lesser extent biological (percolating or trickling) filters. In the latter case, model filters [44, 52] and rotating tubes inclined at a few degrees to the horizontal [35, 60] have been used. Neither of these simulations have yet been considered by the OECD Chemicals Group or the EEC but there is no reason to suppose that the biochemical events taking place in rotating tubes and model filters will be significantly different from those occurring in the more well tried model activated sludge systems.

Neither of the two activated sludge systems recommended by the OECD, and accepted by the EEC, is ideal in reproducing all the conditions operating in full scale plants, but they are serviceable compromises, which retain a "fail-safe" character while giving a reasonable guide to the behaviour of test substances in waste-water treatment. One shortcoming is that neither model permits an anaerobic zone in the settlement tank or elsewhere and this may be important for the degradation of some chemicals. For example, halogenated compounds are reported to be more readily dehalogenated under anaerobic conditions.

The first recommended method, employing the *Husmann Unit* (Fig. 3) and known as the *OECD Confirmatory Test* [46] was originally applied to the assessment of the biodegradability of synthetic surfactants (detergents). It has been adapted for general use with soluble organic chemicals by substitution of DOC analysis for the specific analyses of surfactants. In the standard method [61], synthetic sewage is used to overcome discrepancies associated with the variable nature of real sewage. The sewage is applied to the 3-l aeration vessel at a rate of 1 l/h, giving an average retention time of the liquid phase of 3 h, and the dissolved oxygen concentration is kept at 2 mg/l or over. The mixed liquor passes to the settlement tank of about the same volume from which treated effluent overflows into a collection vessel, prior to analysis for specific chemical or DOC. Settled sludge is returned to the aeration vessel by means of an air-lift pump; the rate of return is so high (because of the geometry of the pump) that the settled sludge

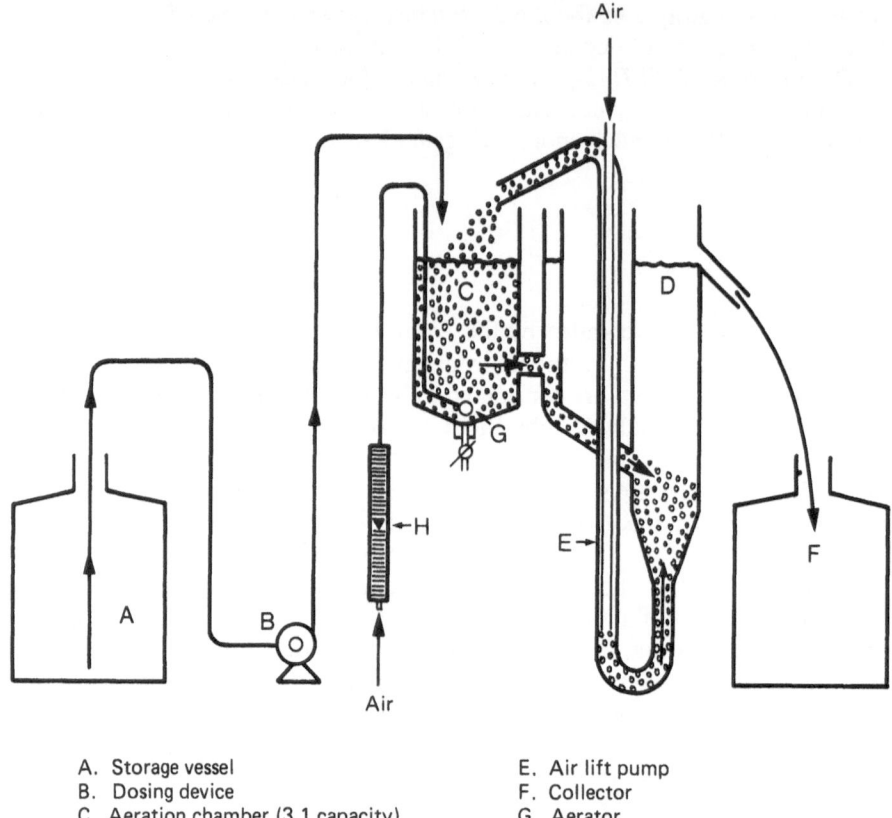

A. Storage vessel   E. Air lift pump
B. Dosing device   F. Collector
C. Aeration chamber (3.1 capacity)   G. Aerator
D. Settling vessel   H. Air flow meter

Fig. 3. Husmann unit (by permission of HMSO, London: Ref. 35)

does not become anaerobic, as it does on the large scale. Sludge is wasted from the system twice weekly to maintain the concentration of suspended solids in the aeration vessel at 2.5 g/l.

An alternative to the Husmann apparatus is the *Porous Pot system* [61, 62] (Fig. 4), which consists of a 3-l aeration vessel of porous polythene, in which the sludge is retained, contained within a non-porous cylinder. Effluent filters through the pores into the annular space and overflows into a collection vessel; there is no settlement of sludge.

The 3-h liquid retention time is shorter than that normally allowed on the full scale (in the UK it is 6–12 h) so that any removal of test substance in the laboratory systems will not over-estimate biodegradation which would occur under normal sewage treatment conditions.

Since effluents from activated sludge units treating sewage alone contain DOC, two units must be operated in any assessment; one would receive synthetic sewage alone, the other sewage plus test substance, which is either added to the sewage daily or pumped direct to the aeration vessel by a separate system. Control

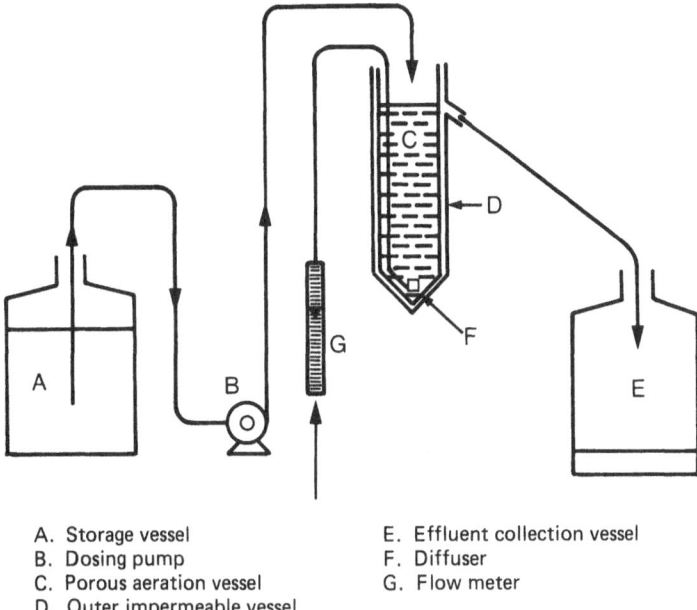

A. Storage vessel
B. Dosing pump
C. Porous aeration vessel
D. Outer impermeable vessel
E. Effluent collection vessel
F. Diffuser
G. Flow meter

Fig. 4. Porous pot unit (by permission of HMSO, London: Ref. 35)

effluents contain 6–10 mg DOC/l, so that, in order to allow for the accuracy and precision of carbon analysers, the concentration of test substance dosed should be at least 10 mg C/l, and preferably 15 mg C/l. In some cases it may be advisable to start with a low concentration and gradually increase to the required level. The removal of test substance is calculated on a daily basis as the difference between the concentrations of DOC in the test and control effluents expressed as a percentage of the concentration of DOC dosed. In the standard test, operation for not more than 6 weeks, if necessary, is stipulated to allow a steady state to be attained and the percentage removal is taken as the mean of at least 15 determinations over three weeks of steady state operation. The time taken to reach the steady state is called the "running-in" or "working-in" period.

This method measures bio-elimination so it may be necessary to distinguish between biodegradation and adsorption of the test substance onto the sludge. Adsorption is usually most marked at the start of the test and generally an equilibrium is reach, so adsorption may be identified by examination of the course of bio-elimination. Biodegradation is usually indicated as a steady upward trend of removal of test substance to a constant "plateau" at the steady state.

Variations. A number of variations have been agreed by the EEC Degradation/Accumulation Sub-Group. Domestic sewage may be substituted for synthetic sewage; this has been shown to give more variable, but still acceptable, values for percentage removal, while significantly increasing the removal of some substances [63]. Also, an alternative method of wasting sludge, based on a kinetic approach, can be used in which 500 ml of the aeration tank contents is discarded daily to give a sludge age, or retention time, of 6 d.

Lastly, the units may be operated in the "coupled" mode [64]. This involves the daily transfer of sludge between aeration vessels of a pair of units; for example, 1.5 l of each sludge is transferred. One of the pair is dosed with sewage plus test substance and the other with sewage alone. In calculating the percentage removal, a small correction must be made for known amounts of test substance transferred to the control vessel. The advantage of the method is that it equalises the population of micro-organisms, quantitatively and qualitatively, in the two vessels and it is reported [64] to give a higher precision in percentage removal values than obtained in the non-coupled mode. However, in an assessment of 5 chemicals of varying ease of biodegradability no significant differences between the two modes was reported [65].

The same sets of apparatus can also be used to assess any effects of a test substance or industrial waste water on the treatment processes. Analyses for BOD, COD, DOC, ammonia-N, and oxidised-N in effluents from units receiving the test substance or waste water and control units will reveal whether inhibition is occurring.

All the OECD test methods are to published in modified form by the EEC.

Choice of Methods: Schemes of Testing

Where biodegradability under a specific set of environmental conditions requires evaluation, the choice of appropriate method may be self-evident. In general, however, most test methods, although providing similar conditions to specific natural environments, must not be regarded as simulation tests. Instead, they should be considered to be tests in which the conditions offer a greater or lesser chance of degradation at the concentration tested. Methods with short test periods and low inocula concentrations offer relatively unfavourable conditions, whereas those with longer test periods, high inocula levels and enrichment procedures offer highly favourable conditions.

Normally in hazard assessments there is time for thorough testing by an agreed set of procedures, but in cases of *accidental spills* and *requests for discharge* of industrial waste waters rapid tests must be applied. A technique [66] which has been found useful as a rough guide is measurement of respiration rate, by means of an oxygen electrode, of endogenous activated sludge alone and with added test substance or waste water. If a manometric respirometer is available, a similar test can be carried out over a period of hours or days, as required.

Before proceeding to discuss schemes of testing in hazard assessment of new chemicals, it is worthwhile comparing the methods available in terms of apparatus required and ease of performing. Three of the indirect methods – BOD, Closed Bottle and Sturm – require only standard, readily available apparatus both for incubation and subsequent analysis. The other indirect method – manometric/ automatic respirometry, MITI, and EEC versions – needs special apparatus for the automatic electrolytic production of oxygen and recording of oxygen uptake. The MITI inoculum is lengthy in preparation, while for the EEC version the inoculum is easily obtained.

Though the direct, die-away methods utilise standard apparatus for incubation, expensive DOC analysers are necessary, together with apparatus for automatic sample presentation to reduce labour effort and costs. The ultrafiltration step requires relatively costly membranes and can be time-consuming. The AFNOR inoculum takes more time to prepare than the other direct methods, but is probably more reproducible.

The present authors have no practical experience of the *RDA method* [55]; it would seem to be more labour intensive than the above screening methods but gives more information.

The incubation apparatus needed for the two *inherent methods* is relatively simple and both require DOC analysers. The attention which has to be given to each type is about the same; the Zahn-Wellens reaction mixture requires daily pH measurement and, perhaps, adjustment, while in the SCAS method the supernatant has to be replaced daily by fresh sewage and test substance.

Simulation methods require a great deal of operator time both for running the units and for analysis. Husman units are more complex and therefore more costly to construct than porous pots, which have the added advantage of being able to be easily installed in water-baths. A disadvantage of porous pots is their tendency to block, requiring periodic replacing and cleaning, while sludge often "bulks," and fails to settle in the Husman units causing it to be lost in a random way from the settlement chamber. If domestic sewage is available its use avoids the need for preparation of synthetic sewage, the components of which have to be bought. Provided that domestic sewage has been properly settled before application, its use does not cause tubing, etc. to become blocked more frequently than when synthetic sewage is employed.

Sparingly soluble substances can be accommodated in most of the above tests provided that 10–15 mg of the material can be dissolved in 1 litre of the previously prepared medium or synthetic sewage. *Insoluble chemicals* may be tested by the respirometric and Sturm methods but no methods of dispersion have yet been agreed.

Volatile chemicals may be assessed by a version of the Closed Bottle test using a special flask [32] which allows the test substance to be injected into the aerated, inoculated medium.

In *hazard assessment of new chemicals* it is in only rare cases sufficient to determine the ratio of BOD_5 to COD, which is the minimum legal requirement. A value of, say, 0.5 or more would indicate a readily biodegradable substance while for values lower than this further testing is required.

The subsequent pattern of testing adopted is determined by the analyst's purpose, but it is advisable at this stage to assess the toxicity of the substance so that a non-inhibitory concentration can be used in the biodegradability assessment. A modified BOD test [66] or respiration test [67, 68] is suitable for this purpose.

A suggested strategy based on that of the OECD Chemicals Group is shown in Fig. 5. It consists of carrying out tests in a series of steps, starting with the relatively simple and inexpensive methods and progressing to the more complex and

H. A. Painter, E. F. King

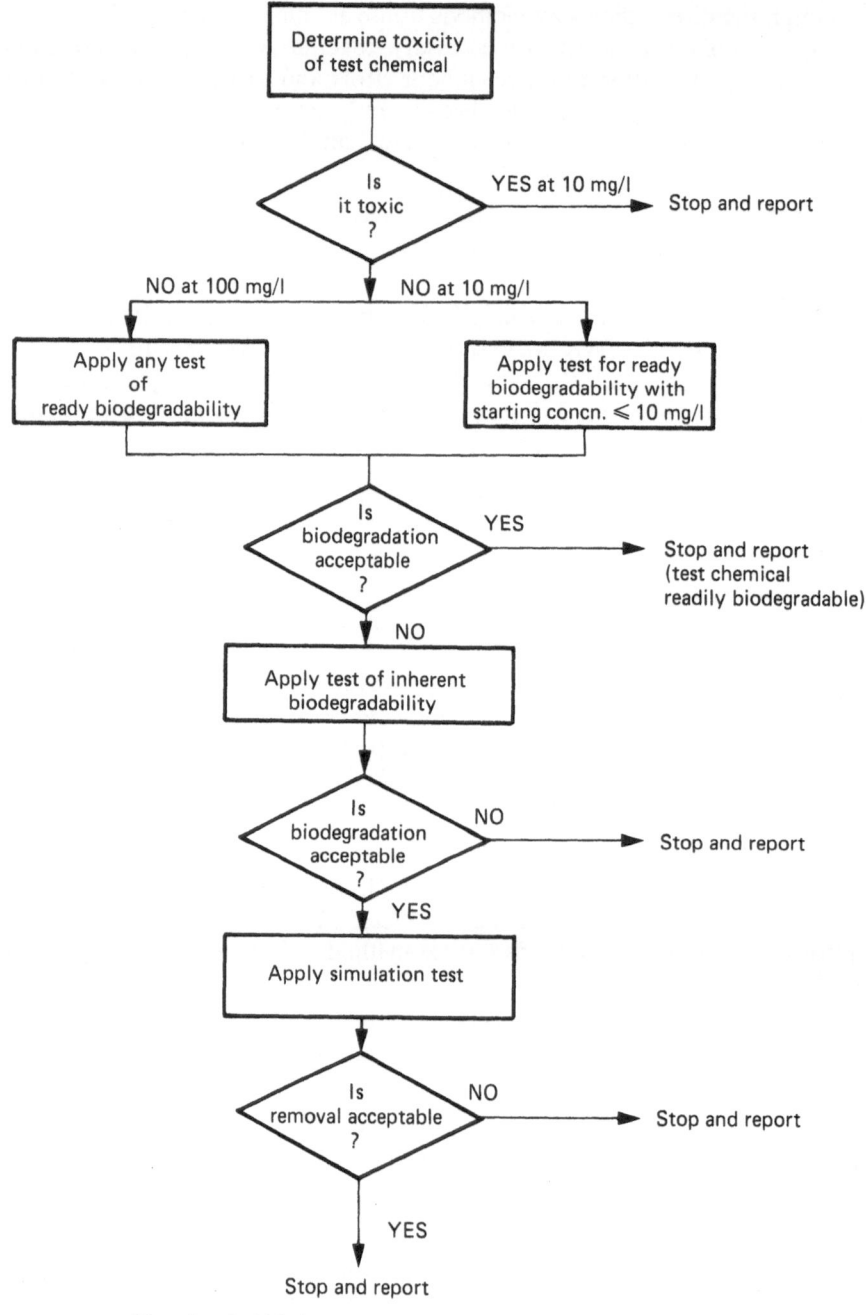

Fig. 5. Strategy for testing biodegradability of chemicals soluble to at least 10 mg/l (by permission of HMSO, London: Ref. 35)

costly ones. The steps are related to the different categories of biodegradability recognised by the OECD. Any of the five screening tests is first applied; if, however, the substance is to be discharged directly to a river it would be prudent to apply the Closed Bottle or modified OECD methods, since they employ low concentrations of micro-organisms. A positive result in any of the screening tests indicates that the substance is readily biodegradable and need not be further examined.

However, a negative result cannot be taken as evidence that the substance is not biodegradable; further tests should be made, perhaps with a lower concentration of test substance if its toxicity had not been already assessed. Alternatively, a test with a higher concentration of inoculum should, if possible, be used or an acclimatised seed could be tried, but the method and period of acclimatisation must be reported. Another variant is to apply the Bunch-Chambers method to test the need for cometabolism. If these fail the Zahn-Wellens or SCAS methods should be applied. Distinctions between results obtained by these two methods have been only rarely investigated, but it is thought that the SCAS method offers a greater chance of inducing a compound to become biodegraded.

For compounds found to be inherently but not readily degraded and to be discharged to sewers, the next step is a simulation test – either the activated sludge or rotating tube methods. (Ideally, for substances to be discharged predominantly to rivers a river simulation test should be carried out.) Since in activated sludge simulation methods the ratio of biomass to test substance is high, significant adsorption onto sludge flocs may have occurred: to be sure that biodegradation has taken place, a separate investigation is required, e.g. analysis for test substance on sludge. In cases of little or no bio-elimination or a negative result, the simulation test could be repeated with a lower concentration of test substance, with a higher retention time of sewage or for a period greater than the normal 9 weeks.

Other Schemes

Many of the new chemicals being produced have novel structures and few are readily biodegradable. It may, therefore, be more rewarding to change the scheme of testing, especially if a large number of chemicals have to be tested simultaneously. Over the last few years studies at the Water Research Centre, UK, on chemicals known to be discharged to sewage treatment works have indicated that it is more cost-effective to adopt a different approach. Of 55 chemicals tested only 11 were readily biodegraded, a further 6 were inherently biodegraded, leaving 38 non-biodegradable. The application of the OECD strategy led to a total of 158 separate tests. The same information could have been obtained by first applying the SCAS test to all 55 substances and then subjecting the 17 (11+6) inherently degradable chemicals to a screening test, which would have revealed 11 to be readily biodegradable. By this scheme the total number of tests would have been only 107, representing a considerable saving of effort.

Interpretation of Results

In attempts to achieve a more nearly uniform interpretation of data from biodegradability tests between and within national and international bodies, the
screening tests have been subjected to scrutiny and validation by various organisations including OECD and EEC. This has been done by exchanging experi-
ences and, more formally, by carrying out calibration exercises ("ring tests"), as
mentioned earlier, in which every aspect has been strictly controlled, including the
source of the inoculum, but for practical reasons not the inoculum itself. Not only
have ring tests brought to light and subsequently remedied technical points, such
as changes in pH value and interfering oxygen uptake by oxidation of ammonium
salts, but more significantly they have built up a body of knowledge and experience which enables the majority of results obtained to be accepted with a high degree of confidence. Along with other investigations of the behaviour of various
chemicals in screening, inherent and simulation tests, the results of ring tests con-
firm that, in general, substances classified as readily biodegradable will indeed degrade rapidly and completely in the environment. Substances classified as only inherently degradable range from those which will degrade completely in the environment after a period (of acclimatisation) to those which degrade very slowly
or not at all under environmental conditions. Their fate can be decided only by
application of a simulation test.

However, most data are not published in easily accessible places or remain un-
published in the files of various organisations. Data on new chemicals presented
to competent authorities under various laws (see next section) are not published
for reasons of confidentiality, so that few additional results are likely to appear.

The above investigations have revealed a minority of chemicals which behave
erratically in screening tests, even after allowing for experimental error. They degrade in some laboratories but not in others; they degrade on some occasions in
one laboratory but not on others; and some even degrade in some replicates and
not in others on the same occasion. Such disparate results have been obtained
with compounds such as hexamethylenetetramine, diethylene glycol, 4-nitro-
phenol, and pentaerythritol. These discrepancies have come to light because the
substances mentioned have been tested many times in many laboratories, but the
situation with new chemicals is different. New chemicals have to be reported to
a competent authority but are not examined in anything like such an intensive
manner, so that any erratic behaviour may often pass undetected. The reasons for
such behaviour are not established; various explanations are possible – the competent bacteria are not widely distributed and few in number, they may be excep-
tionally nutrionally fastidious; or some species may be readily inhibited by the
substrate. It is thus advisable to repeat a test or to apply a different screening
method if disparate replication of results are obtained on the first attempt. In the
absence of firm information, these compounds would have to be assumed to be
non-readily degradable for purposes of hazard assessment.

Another phenomenon, important for insoluble compounds, is the frequently
reported percentage of the theoretical oxygen uptake of less than the recommended 60%. Simultaneous determination of the percentage DOC removed has
often shown complete removal (>90%), while the oxygen uptake was found to

be as low as 45%. These low values are probably due to more carbon being channelled to cell synthesis though the reason is not clear.

Legislation

The absorption of scientific and technical information into relevant legislation is always difficult and hazard assessment is no exception. Data on biodegradation, along with other information, have to be presented in dossiers to competent authorities (in the EEC, to member states [69]). Results are more acceptable if recognised protocols and methods are used, carried out under some system of good laboratory practice. Deviations from the recognised methods, as well as other methods, are acceptable, provided that they are scientifically justified. In general the EEC has adopted the OECD methods and, with modifications, the scheme of testing.

For a new chemical to be successfully put on the market it does not have to pass a biodegradation test, since many other properties (e.g. toxicity to fish, alga, etc.) and its likely yearly production are taken into account. Thus, the OECD scheme although used as a guide, is not followed strictly; for example, in some cases a simulation test may be demanded following a negative result in a screening test, omitting a test for inherent degradation, if the safety margin between the probable environmental concentration and the maximum accepted concentration (based on toxicity to fish or other aquatic organisms) is too small.

Future Work

The greatest need in biodegradability testing is for a batch method(s) which will give results capable of being used to make accurate predictions of what will happen in the environment. This will involve being able to obtain from batch tests characteristics of the biodegradation rate which are sufficiently independent of the conditions of the method to allow their use in calculating degradation rates in waste water treatment plants, in rivers, in the sea, and in soil. These characteristics would obviously be the specific growth rate, μ, and the saturation constant (or some function of these) of the bacteria able to degrade the substance either directly or after acclimatisation. If acclimatisation is to be used, the relationship between duration and other conditions of acclimatisation, on the one hand, and the subsequent acquisition of degradative ability requires study so that conditions may be chosen to allow accurate prediction of behaviour in simulation tests. "Over"-acclimatisation of inocula could result in "false positive" results.

The kinetic equations, involving μ, may be used to extrapolate to rates of removal at environmentally realistic concentrations at the μg/l level, unless other mechanisms become rate-limiting, as they probably are, e.g. rate of entry of substrate into the cell. Practical rates can be determined only by use of ^{14}C-labelled substances.

A promising method is the modification by Blok and Booy [27] of Blok's original repetitive die-away method. Approximate values of μ can be obtained from the increase in percentage removal (or of oxygen uptake) between weekly additions of test substance to media contained in partially filled BOD bottles.

In similar vein maximum specific growth rates have been determined by Birch [70] over a range of temperatures but using a simulation test – a modified porous pot system allowing continuous wastage of sludge. The units are run at various wastage rates until "wash-out" of the substrate occurs at which point the wastage rate is equivalent to the maximum specific growth rate. Both these methods merit further attention.

Other problems needing attention are the establishment of agreed test methods for

 (i) river simulation,
 (ii) degradation under marine conditions,
(iii) degration in the soil,
(iv) anaerobic degradation in sludge, mud etc.

More minor problems – minor because of possible lack of relevance – are the assessment of aerobic degradation of insoluble and volatile chemicals. Finally, hand-in-hand with practical studies should go investigations into the possibility of establishing quantitative relationships (QSBR) between molecular structure and ease of biodegradation, not only because of the intrinsic interest in them, but also as an aid in hazard assessment and saving in effort. At this stage, any QSBR's other than those covering small groups of similar compounds seem a long way off.

References

1. Degradation of synthetic organic molecules in the biosphere, 1972. Available from: Printing and Publishing Office, National Academy of Science, 2102 Constitution Avenue, N.W., Washington, D.C. 20418
2. Gibson, D.T.: In the handbook of environmental chemistry. Vol. 2, Part A (O. Hutzinger, ed.) pp. 161–192. Berlin: Springer-Verlag 1980
3. Gale, E.F.: The chemical activities of bacteria. New York: Academic Press 1952
4. Alexander, M.: Adv. Appl Microbiol. 7, 35 (1965)
5. Horvath, R.S.: Bact. Rev. 36, 146 (1972)
6. Alexander, M.: In: Microbial degradation of pollutants in marine environments (Bourquin, A.W., Pritchard, P.H. eds.) pp. 67–75. EPA, Gulf Breeze 1979
7. Pike, E.B., Curds, C.R.: Soc. Appl. Bact. Symp. 1, 123 (1971)
8. Hawkes, H.A.: In: The oil industry and microbial ecosystems (Chater, K.W.A., Somerville, H.J. eds.) pp. 217–233. London: Heyden and Son 1978
9. Slater, J.H.: In: The oil industry and microbial ecosystems (Chater, K.W.A., Somerville, H.J. eds.) pp. 137–154. London: Heyden and Son 1978
10. Horowitz, N.H.: In: Evolving genes and proteins (Bryson: V, Vogel, H.J. eds.). New York: Academic Press 1965
11. Wu, T.T., Lin, E.C.C., Tanaka, S.: J. Bact. 96, 447 (1968)
12. Slater, J.H., Somerville, H.J.: In: Microbial technology: Current state, future prospects (Bull, A.T., Ellwood, D.C., Ratledge, C. eds.) pp. 221–262. Cambridge: Cambridge Univ. Press 1979
13. Various authors in microbial degradation of xenobiotics and reclacitrant compounds (Leisinger, T., Hütter, R., Cook, A.M., Nuesch, J., eds.). London: Academic Press 1981
14. Neilson, A.H., Allard, A.-S., Remberger, M.: Biodegradation and transformation of recalcitrant compounds, In: The handbook of environmental chemistry. Vol. 2, Part C (Hutzinger, O. ed.) pp 29–86. Berlin: Springer-Verlag 1985
15. Painter, H.A.: Proc. R. Soc. Lond. B 185, 149 (1974)
16. Yonezawa, Y., Urushigawa, Y.: Chemosphere 3, 139 (1979)

17. Paris, D.F., Steen, W.C., Baugham, G.L., Barnett, J.T.: Appl. Envir. Microbiol. *41*, 603 (1981)
18. Enslein, K. et al.: Literature study on the biodegradability of chemicals in water, U.S. Environmental Protection Agency, Cincinnati, Ohio 1981
19. Mudder, T.I.: Development of empirical structure biodegradability relationships. Thesis, Univ. of Iowa 1981
20. Pitter, P.: Wat. Res. *10*, 231 (1976)
21. Painter, H.A., Denton, R., Quarmby, C.: Wat. Res. *2*, 427 (1968)
22. Reiner, E.A., Chu, J.P., Kirsch, E.J.: In: Pentachlorophenol: chemistry, pharmacology and environmental toxicology (Rao, K.R. ed.) pp. 67–82. New York: Plenum Press 1978
23. Jones, G.L.: Wat. Res. *7*, 1475 (1973)
24. Boethling, R.S., Alexander, M.: Environ. Sci. Tech. *13*, 989 (1979)
25. Paris, D.F., Steen, W.C., Burns, L.A.: In: The handbook of environmental chemistry, Vol 2/Part B, pp. 73–81 (Hutzinger, O. ed.). Heidelberg: Springer-Verlag 1982
26. Painter, H.A., King, E.F.: Reg. Toxicol. Pharmacol. *3*, 144 (1983)
27. Blok, J., Booy, M.: Ecotoxicol. Environ. Safety (in press)
28. Larson, R.J.: Residue reviews *85*, 159 (1983)
29. Howard, P.H. et al.: Review and evaluation of available techniques for determining persistence and routes of degradation of chemical substances in the environment. U.S. Environmental Protection Agency, available from National Technical Information Service. Springfield, Virginia, 22151
30. Gilbert, P.A., Watson, G.K.: Tenside detergents *14*, 171 (1977)
31. de Kreuk, J.F., Hansveit, A.O.: In: The oil industry and microbial ecosystems (Chater, K.W.A., Somerville, H.J. eds.) pp. 155–170. London: Heyden and Son 1978
32. de Kreuk, J.F., Hansveit, A.O.: In: Degradability, ecotoxicity, and bioaccumulation, pp. 111–147. Government Publishing Office. The Hague, Netherlands 1980
33. Bock, K.J., Stache, H.: In: The handbook of environmental chemistry, vol. 3. Part B (Hutzinger, O. ed.) pp 164–199. Heidelberg: Springer-Verlag 1982
34. OECD Report on chemicals testing programme degradation/accumulation (final report). Berlin and Tokyo, Dec 1979, also 1980
35. Methods for the examination of waters and associated materials, assessment of biodegradability. London: Her Majestry's Stationery Office, 1983
36. Cook, A.M., Hütter, R.: In: Microbial degradation of xenobiotics and recalcitrant compounds, pp. 237–249 (Leisinger, T., Cook, A.M., Hütter, R., Nüesch H. eds.). London: Academic Press 1981
37. Gerike, P., Wierich, P.: In: Principles for the interpretation of the results of testing procedures in ecotoxicology, pp. 128–142. Commission of the European Communities, Luxembourg 1982
38. Payne, W.J., Wiebe, W.J., Christian, R.R.: Bioscience *20*, 862 (1970)
39. Association Francaise de Normalisation, Method for the evaluation in aqueous medium of the biodegradability of total organic products, T90/302 (1977)
40. The biodegradability and bioaccumulation of new and existing chemical substances. Ministry of International Trade and Industry, Japan 1978
41. Sturm, R.N.: J. Amer. Oil Chem. Soc. *50*, 159 (1973)
42. Zahn, R., Wellens, H.: Z. Wasser Abwasser-Forsch. *13*, 1 (1980)
43. Zahn, R., Wellens, H.: Chemiker. Z. *98*, 228 (1974)
44. Gerike, P., Fischer, W.K.: Ecotoxicol. Env. Safety *5*, 45 (1981)
45. Fischer, W.K.: Fette-Seifen-Anstrichmittel, *65*, 37 (1963)
46. OECD Proposed method for the determination of the biodegradability of surfactants used in synthetic detergents. Paris (1976)
47. Council Directives 73/404 and 73/405. Official Journal of the European Communities, No. L347/53 7th Dec 1973
48. Council directive, amending 73/405. Official Journal of the European Communities, No. C112/4, 14th May 1981
49. Council directive, amending 73/404. Official Journal of the European Communities, No. L109/1, 22nd April 1982
50. OECD Guidelines for testing chemicals, OECD, Paris 1981

51. International organisation for standardisation, ISO/DIS 7827 (1983)
52. Gerike, P., Fischer, W.K.: Ecotoxicol. Env. Safety *3*, 159 (1979)
53. Kitano, M.: Comparison of four OECD ready biodegradability test methods. Presented at SECOTOX Meeting, Neuherberg, Oct 1983
54. Bunch, R.L., Chambers, C.W.: J. Wat. Poll. Cont. Fed. *39*, 187 (1967)
55. Blok, J.: Int. Biodeterioration Bull. *15*, 57 (1979)
56. OECD Updating Panel: Inherent biodegradability: Zahn-Wellens/EMPA Test, 302B, Feb 1984
57. Schefer, W.: Forum-Stadte-Hygiene 110 (1978)
58. Soap and detergent association. Biodegradation Subcommittee. J. Amer. Oil Chem. Soc. *46*, 432 (1969)
59. Proposed method C120/79/GER in OECD chemicals testing programme, Vol. II Part 3. Berlin and Tokyo 1979
60. Tomlinson, T.G., Snaddon, D.H.M.: Int. J. Air Wat. Poll. Control *10*, 865 (1966)
61. Biotic degradation: Activated sludge simulation tests, DGXI/605/82, EEC (to be published)
62. Painter, H.A., King, E.F.: Technical report 70. Water Research Centre, UK (1978)
63. King, E.F., Painter, H.A., Solbé, J.F. de L.G.: Report to the department of the environment, 612-M. Water Research Centre, UK (1984)
64. Fischer, W.K., Gerike, P., Holtmann, W.: Wat. Res. *9*, 1131 (1975)
65. Painter, H.A., King, E.F., Solbé, J.F. de L.G.: Report to the department of the environment, 730-M. Water Research Centre, UK (1984)
66. Methods for the examination of water and associated materials, methods for assessing the treatability and toxicity of chemicals. London: Her Majestry's Stationery Office 1984
67. OECD updating panel: Activated sludge respiration, inhibition test, draft ET82.6 (to be published)
68. International organisation for standardisation, test for inhibition of oxygen consumption by activated sludge, DP8192 (1983)
69. Council directive 67/548, 6th amendment. Official Journal of the European Communities, no. L259/10, 15th Oct 1979
70. Birch, R.R.: Biodegradation of nonionic surfactants. Presented at SECOTOX Meeting, Neuherberg, Oct 1983

The Fugacity Concept in Environmental Modelling

S. Paterson, D. Mackay

Department of Chemical Engineering and Applied Chemistry
University of Toronto
Toronto, Ontario, Canada M5S 1A4

Introduction

The Fugacity Concept

Recently, the fugacity concept has been introduced and used to model the fate of chemicals which may be accidentally or deliberately released into the environment [1–3]. Its application to environmental fate determinations of toxic substances results in simplified equations for partitioning, transport, and reaction processes and their assembly into a coherent model. Expressing the distribution of contaminants in the environment in terms of fugacity rather than concentration facilitates interpretation of the dynamic processes to which the substances are subject.

In this chapter, the thermodynamic basis of fugacity, especially its relation to concentration, is reviewed briefly. Two types of fugacity calculations are then described, a series applying to evaluative environments or units worlds, and those for real environments such as lakes. Finally, the role of fugacity calculations in assessing human exposure is discussed.

Fugacity, derived from the Latin word fugere – "to flee," may be defined as the escaping tendency of a substance from a phase. It is an expression of "activity" and as such has been applied mainly to thermodynamic problems involving phase equilibria, especially to calculations encountered in chemical separation processes such as liquid extraction, distillation, and absorption.

Partitioning of a small amount of substance (such as benzene) between two phases (such as water and air) at constant temperature and pressure, results in a relatively constant ratio of concentrations between these phases. This may be expressed as the Nernst Distribution Law, or Henry's Law (the former generally being applied to two liquid phases and the latter to gas and liquid phases). A distribution or partition coefficient K_{12} between the two phases (1 and 2) may be defined as the ratio of concentrations C_1/C_2. Such coefficients are widely used in environmental science to describe equilibria between air, water, soils, sediments, and biota such as fish. The coefficient is primarily a function of temperature and the chemical natures of the substance and the two phases. The relative natures of the chemical interactions between the substance and the two phases determine the partition coefficient.

Gibbs described such partitioning between phases in terms of chemical potential μ. He showed that the criterion of equilibrium for a substance distributed between two phases is that its chemical potential in each phase is equal. This concept of chemical potential is difficult to grasp and cumbersome to use mathematically since it is logarithmically related to concentration.

The concept of fugacity was later introduced by Lewis [4] as an alternative criterion of equilibrium which is easier to grasp since it is more readily visualized. It is usually linearly related to concentration at the low concentrations which prevail in the environment. It has units of pressure.

The definition of fugacity involves two statements, the first establishing its differential relationship to chemical potential, and the second its absolute value defined for a gas phase, namely,

$$d\mu = RTd\ln f \tag{1}$$

and

$$f/P \rightarrow 1.0 \text{ as } P \rightarrow O, \tag{2}$$

where R is the gas constant (8.314 J/mol K), T is absolute temperature (K), f is fugacity (Pa), and P is pressure (Pa).

The form of the differential equation was suggested by the relationship between μ and pressure for an ideal gas at constant temperature since

$$d\mu = vdP = (RT/P)dP = RTd\ln P, \tag{3}$$

where v is molar volume (m^3/mol). Fugacity is thus equivalent to pressure or partial pressure for an ideal gas. The second definition requires that at low pressure when ideal conditions prevail, f and P become equal. Equilibrium is thus attained in the air-water-benzene example when the escaping tendency of the benzene from air to water is exactly balanced by the opposing escaping tendency from water to air.

The term fugacity instead of partial pressure can be applied to phases such as water, soils or even biota without the conceptual difficulty of considering partial pressures to exist in these non-gaseous phases.

More detailed derivations of the fugacity concept can be found in texts by Prausnitz [5] or Van Ness and Abbott [6]. It is not, however, necessary to understand the thermodynamic basis of fugacity in order to use it. The essential point is that fugacity is a potential quantity which characterises equilibrium partitioning of mass, just as temperature characterises partitioning of heat. Just as heat flows from high to low temperature, mass diffuses from high to low fugacity, and there is no net diffusion when equal fugacities are reached.

Fugacity Capacities (Z)

The coefficient which relates fugacity to concentration is termed the fugacity capacity Z ($mol/m^3 \cdot Pa$) and may be considered analogous to a heat capacity. The linear relationship is given by

$$C = fz.$$

Each chemical has a unique value of Z for each phase at a defined temperature. The Z values thus depend on the environmental temperature, the physical chemical properties of the substance and the nature of each phase into which it partitions. The physical properties that determine the fugacity capacity are primarily vapor pressure, aqueous solubility and octanol-water partition coefficient (K_{ow}). Some of these properties may be estimated; for example, correlations exist between vapor pressure and boiling point [7], and between solubility, K_{ow} and melting point [8]. Definitions of Z values are given in Table 1 and their relationship to partition coefficients in Figure 1.

It is important to recognize that a Z value is equivalent to "half" of a partition coefficient, i.e. a partition coefficient is merely the ratio of two Z values.

$$K_{12} = \frac{C_1}{C_2} = \frac{Z_1 f}{Z_2 f} = \frac{Z_1}{Z_2}.$$

Table 1. Definition of Z values and illustrative values for Mirex

			Mirex values
Molecular weight	MW	g/mol	545.6
Vapor pressure	P^s	Pa	1.33×10^{-4}
Aqueous solubility	C^s	mol/m^3	1.28×10^{-7}
Octanol-water partition coeff.	K_{ow}		7.76×10^6
Bioconcentration factor	K_B		3.72×10^5
(correlation $K_B = 0.048\ K_{ow}$)[9]			
Organic carbon partition coeff.	K_{oc}		3.19×10^6
(correlation $K_{oc} = 0.411\ K_{ow}$)[10]			
Z_1 Air	1/RT		4.03×10^{-4}
Z_2 Water	C^s/P^s		9.64×10^{-4}
Z_3 Soil	$Z_2 K_{oc} \phi_3 \varrho_3$		92.3
ϕ = fraction organic content = 0.02			
ϱ = density = 1.5 kg/L			
Z_4 Bottom sediment	$Z_2 K_{oc} \phi_4 \varrho_4$		184.7
ϕ = fraction organic content = 0.04			
ϱ = density = 1.5 kg/L			
Z_5 Suspended sediment	$Z_2 K_{oc} \phi_5 \varrho_5$		184.7
ϕ = fraction organic content = 0.04			
ϱ = density = 1.5 kg/L			
Z_6 Biota (fish)	$Z_2 K_B \varrho_6$		359
ϱ = density = 1.0 kg/L			
v = molar volume (m^3/mol)	T = 298 K		R = 8.314 J/mol K

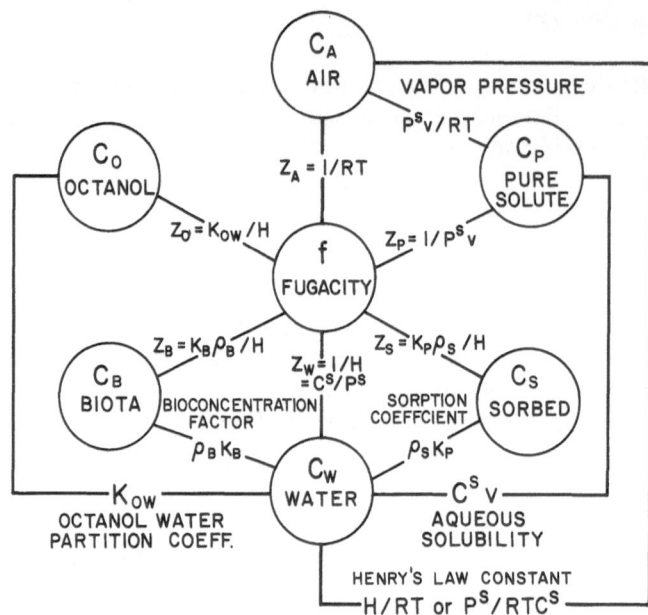

Fig. 1. Relationships between fugacity capacities and partition coefficients. Symbols are defined in Table 1

The advantage of using Z values instead of K values is that only one Z is defined per compound per phase; thus in a six phase calculation there are six Z values, whereas there are potentially 30 K values. A partition coefficient contains information about two phases; thus it may be difficult to determine the relative contributions of each phase to variation in K. As is shown later, the use of Z values simplifies the equations describing interphase transport.

Evaluative and Real Environments

Real environments are usually complex and their properties and process rates are rarely adequately understood. Chemical concentrations vary temporally and spatially. The bulk movements of air, water, suspended solids, and biota are irregular and difficult to describe numerically. Reactions are numerous, interactive and vary with time and space. The effort necessary to describe these complexities may detract from the interpretation of the broad patterns of behavior of the chemical as distinct from the behavior of the environment.

    The evaluative approach presents a simplified model of the environment without attempting to simulate it. The use of such model environments for elucidation of the environmental behaviour of chemicals was first suggested by Baughman and Lassiter [11] in 1978. This led to the EXAMS model [12], the studies of selected chemicals by Smith et al. [13], the development of "Unit Worlds" by Neely and Mackay [14] and Mackay and Paterson [2], and the incorporation of similar

Unit Worlds into hazard assessment by Schmidt-Bleek et al. [15] and Hushon et al. [16]. The evaluative approach eliminates concerns about environmental identification and allows a general behavior pattern to be determined. However, this makes direct validation impossible. The credibility of conclusions based on evaluative environments may be reinforced by demonstrating that they can be applied successfully to microcosms, to well controlled outdoor environments such as small ponds or agricultural plots, or to rivers, lakes, and estuaries.

In summary, modeling of evaluative and real environments should be viewed as complementary. Evaluative models are suitable for assessment of new chemicals, for comparing chemicals, and for obtaining general chemical fate patterns. Real models are needed to elucidate the actual or potential nature of specific contamination situations and remedial actions. The use of similar or identical calculation techniques in each case is essential if credibility is to be achieved for the evaluative case.

One example of an evaluative environment is the "Unit World" which has been used extensively by Mackay and co-workers [2]. It consists of six compartments each with a corresponding number – air (1), water (2), soil (3), bottom sediment (4), suspended sediment (5), and biota (6). Within each compartment, the medium is assumed to be well-mixed and the concentration of the chemical homogeneous. The area is arbitrarily defined as 1 km square with an atmosphere 6 km high. This height effectively corresponds to the quantity of air extending to an altitude in the real environment of 10 km, where diffusion from the troposphere to stratosphere occurs. The 6 km figure results in an air volume of 6×10^9 m^3. The water column has an average depth of 10 m and covers 70% of the area, as suggested by the percentage of water on the Earth's surface. The volume of water is, therefore, 7×10^6 m^3 and is assumed to be fresh.

The active layer of sediment is assumed to be 3 cm deep and overlies the inactive or inaccessible deep sediments. Its volume is thus 21,000 m^3.

The terrestrial or soil compartment has an area of 3×10^5 m^3 and a depth of 15 cm. This depth is based on that of a furrow slice in a cultivated field, below which most (but not all) chemicals will not penetrate, in a short time. Its volume is thus 45,000 m^3.

The suspended sediments in the water are assumed to be present at a concentration of 5 cm^3/m^3 of water and thus have a volume of 35 m^3.

For evaluating bioconcentration effects the biota in the water is assumed to be fish at a volume fraction in water of 10^{-6}, i.e., a total volume of 7 m^3.

The unit world and its dimensions are illustrated in Figure 2.

Illustration

To illustrate the use of the fugacity concept for calculating the environmental behaviour of a contaminant we consider the contaminant mirex (a perchlorinated C_{12} hydrocarbon). Mirex is a persistent organochlorine pesticide which was widely used in North America and has been emitted into the environment both in effluents from industrial synthesis and from direct use. Its properties are given in Table 1, as are its deduced Z values.

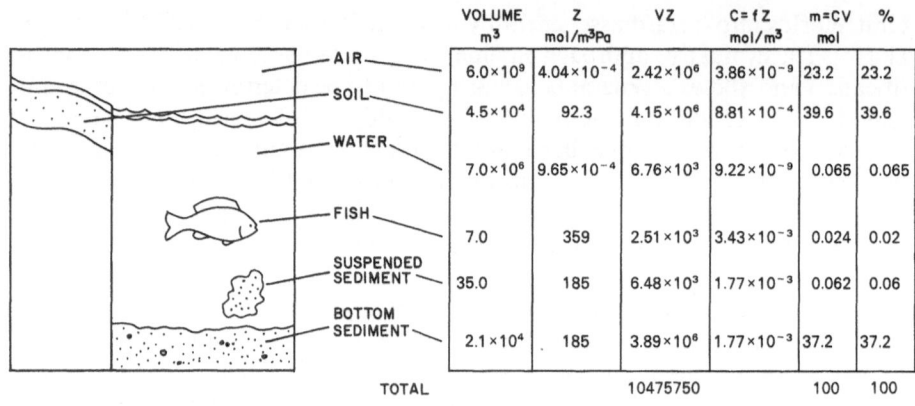

	VOLUME m³	Z mol/m³Pa	VZ	C= fZ mol/m³	m=CV mol	%
AIR	6.0×10^9	4.04×10^{-4}	2.42×10^6	3.86×10^{-9}	23.2	23.2
SOIL	4.5×10^4	92.3	4.15×10^6	8.81×10^{-4}	39.6	39.6
WATER	7.0×10^6	9.65×10^{-4}	6.76×10^3	9.22×10^{-9}	0.065	0.065
FISH	7.0	359	2.51×10^3	3.43×10^{-3}	0.024	0.02
SUSPENDED SEDIMENT	35.0	185	6.48×10^3	1.77×10^{-3}	0.062	0.06
BOTTOM SEDIMENT	2.1×10^4	185	3.89×10^6	1.77×10^{-3}	37.2	37.2
TOTAL			10475750		100	100

$$f = M/\Sigma VZ = 100/10475750 = 9.55\times10^{-6}$$

Fig. 2. Illustration of unit world and level I calculation for Mirex

Fugacity Models of Evaluative Environments

Level I Calculation

A Level I calculation is the simplest form of the model and calculates the equilibrium partitioning of a fixed amount of chemical between the phases. It assumes that each compartment is well-mixed and the chemical is nonreactive.

At equilibrium

$$f_1 = f_2 = f_3 \ldots = f_i,$$

where f_i is the fugacity (Pa) and 1, 2 ... i represent the environmental compartments. A total amount M (mol) of chemical is introduced into the system.

M is known and

$$M = \Sigma\, m_i = \Sigma C_i V_i = \Sigma\, f_i Z_i V_i = f\, \Sigma\, Z_i V_i,$$

from which the only unknown, f, may be calculated as

$$f = M/\Sigma\, Z_i V_i,$$

where m_i (mol) and V_i (m³) are the amounts and volumes for each phase. The phase concentrations C_i (mol/m³) can be calculated as

$$C_i = f Z_i.$$

The amount in each phase is given by

$$m_i = f V_i Z_i = C_i V_i.$$

This level of calculation provides information for each phase on
i) relative concentrations,
ii) relative mass distribution.

The absolute concentration and amount in each phase are arbitrarily determined by the selection of volumes and the total amount in the system.

Figure 2 gives the results of a Level I calculation for mirex. It is distributed mainly between soil, sediment and air with small amounts in the other phases. However, the highest concentration is in the biota and sediment due to the chemical's low solubility and high K_{ow} which result in high sorption and bioconcentration factors.

Level II Calculation

The Level II calculation also assumes equilibrium but has the capability of including reaction or transformation, and, if desired, advection. Transfer between phases is assumed to be rapid compared to reaction and therefore emissions are considered to be into the system as a whole. Reactions can include photolysis, hydrolysis, biodegration, oxidation, or "other," which can represent reactions with other compounds that may be present. These processes are considered to be first-order in the substances, with units of reciprocal time and follow the rate law

$$\text{rate (mol/m}^3 \cdot \text{h)} = kC,$$

where k is the rate constant (h^{-1}).

These first order degradation rates are additive and the total removal rate (mol/h) from a compartment is

$$V_i C_i \Sigma k_j = V_i C_i \, k_i \text{ mol/h},$$

where

$$k_i = \Sigma \, k_j.$$

A mass balance of reaction summed over all compartments with the total input rate of E (mol/h) gives

$$E = \Sigma \, V_i C_i k_i.$$

As in Level I, a common fugacity prevails throughout the system. Substituting fZ_i for C_i yields

$$E = f\Sigma V_i Z_i k_i$$

and

$$f = \frac{E}{\Sigma V_i Z_i k_i}.$$

The concentrations in each phase can be calculated as fZ_i, the amounts m_i as $C_i V_i$, the total amount M as Σm_i and the reaction rates as $C_i k_i$ mol/m$^3 \cdot$ h) or $V_i C_i k_i$ mol/h. The dominant degradation processes may then be ascertained as occurring where $V_i Z_i k_i$ is largest. This is not always intuitively obvious as a high rate constant may combine with a low Z or V value to produce an environmentally insignificant reaction rate.

The overall environmental rate constant for removal may be calculated as

$$k_R = \frac{\text{total reaction rate}}{\text{total amount in system}}$$

$$= \Sigma V_i C_i k_i / \Sigma V_i C_i = E/M$$

and the residence time in the environment due to reaction is

$$\tau_R = M/E = \Sigma V_i C_i / \Sigma V_i C_i k_i = 1/k_R ,$$

This reaction residence time can also be termed "persistence."

In many cases advective flow (usually in air or water) plays an important role in chemical fate. If desired, it can be introduced in the form of a first order rate constant k_{Ai} which is equal to G_i/V_i where G_i m³/h is the flow rate into and out of the phase. The phase residence time is thus V_i/G_i hours. If C_{Bi} is the inflow concentration, the mass balance equation for Level II becomes

$$E + \Sigma G_i C_{Bi} = \Sigma V_i C_i (k_i + k_{Ai})$$

and

$$f = \frac{E + \Sigma G_i C_{Bi}}{\Sigma V_i Z_i (k_i + k_{Ai})} ,$$

The phase concentrations, amounts and residence time can be calculated as before.

If there is no reaction and only steady state advection in and out of the system

$$E + \Sigma G_i C_{Bi} = \Sigma G_i C_i = \Sigma V_i k_{Ai} C_i = f \Sigma V_i Z_i k_{Ai} ,$$

The rate constant k_A and residence time τ_A for the situation in which only advection occurs are

$$k_A = \Sigma V_i C_i k_{Ai} / \Sigma V_i C_i = 1/\tau_A \quad \text{and} \quad \tau_A = \Sigma V_i C_i / \Sigma V_i C_i k_{Ai} .$$

If both reaction and advection occur the average overall residence time τ_0 and the overall rate constant k_0 are

$$\tau_0 = M/(E + \Sigma G_i C_{Bi}) = \Sigma V_i C_i / \Sigma (V_i C_i (k_i + k_{Ai}))$$

and

$$k_0 = \Sigma (V_i C_i (k_i + k_{Ai})) / \Sigma V_i C_i = k_R + k_A$$

Thus the overall residence time may be expressed as

$$\frac{1}{\tau_0} = \frac{1}{\tau_R} + \frac{1}{\tau_A} .$$

It can be seen that the rate constants for reaction and advection are additive while it is the reciprocals of residence time which must be used for determining the overall environmental persistence. The overall residence time is always less than or equal to the sum of the reaction and advection times.

Two processes which effectively remove the chemical from the Unit World and which may also be considered to be advection can be included as pseudoreac-

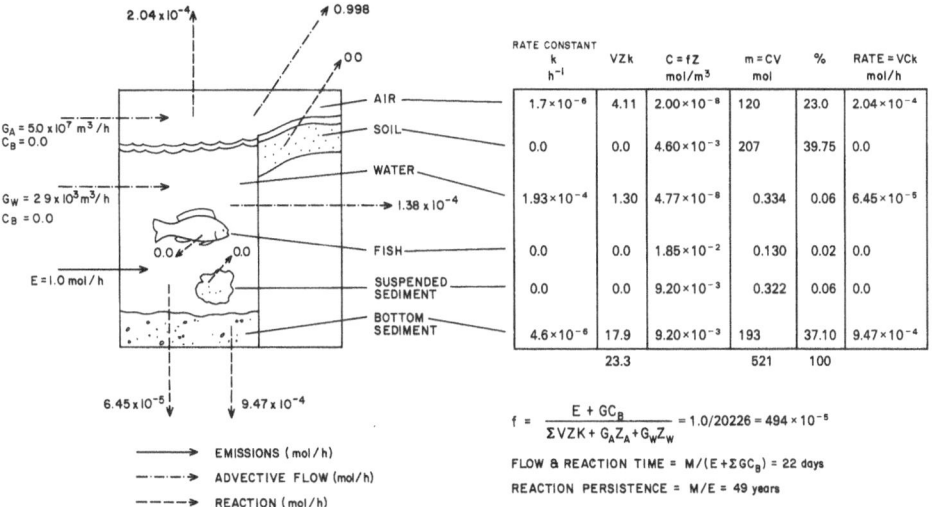

Fig. 3. Level II calculation for Mirex

tions. These are transport from the troposphere to the stratosphere which occurs with a rate constant of approximately 1.7×10^{-6} h^{-1} and sediment burial to a depth greater than 3 cm which typically occurs with a rate constant of 4.6×10^{-6} h^{-1} [14]. These processes apply equally to all chemicals.

Assuming, (purely for illustrative purposes) reaction rate constants for photolysis and oxidation in water of 1.67×10^{-4} and 2.6×10^{-5} (h^{-1}) respectively, and including those for transfer to stratosphere and sediment burial, a Level II calculation can be done for mirex as shown in Figure 3. The emission rate into the unit world is arbitrarily set at 1.0 mol/h together with residence times (V_i/G_i) of 5 days for air and 100 days for water.

The mass percentage distribution is the same as in Level I and it can be seen that sediment burial is the major process of removal. The reaction persistence is extremely long being approximately 49 years. However, air and water advection combined with the low reaction rate in water reduce this residence time in the Unit World to approximately 22 days. Mirex is clearly a very persistent chemical, but it may be rapidly removed from certain localities.

Level III Fugacity Model

Levels I and II are limited in their application since they only consider equilibrium conditions. The Level III calculation introduces interphase transfer resistances in a non-equilibrium, steady state system. This model allows for different fugacities in each phase. It is now necessary to establish into which compartment(s) the emissions take place.

The flux N (mol/h) for diffusive transfer between phases i and j is related to the transfer coefficient or "conductivity" D_{ij} (mol/m$^3 \cdot$ h) by

$$N = D_{ij} f_i - D_{ji} f_j$$

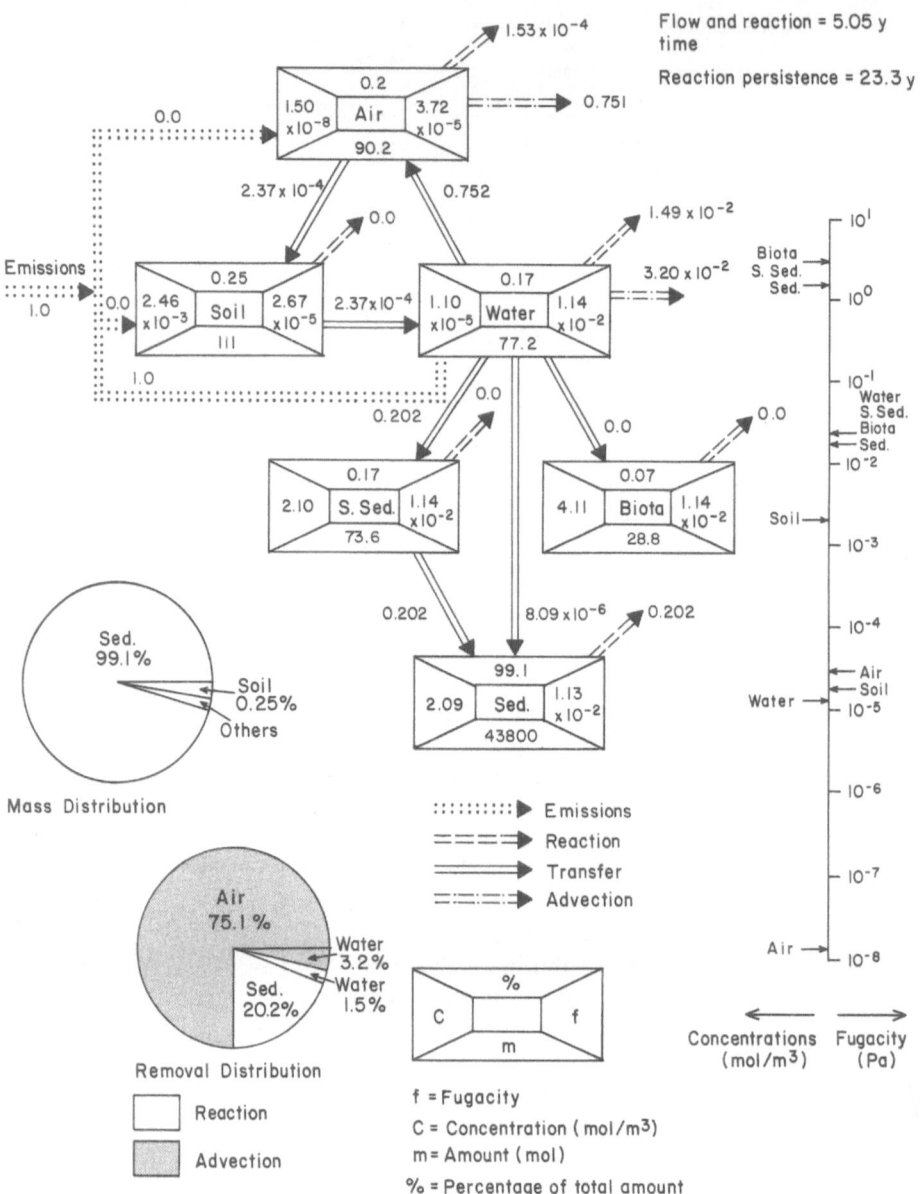

Fig. 4. Level III calculation for Mirex

or for reversible diffusive processes in which D_{ij} and D_{ji} are equal

$$N = D_{ij}(f_i - f_j).$$

These D values can be calculated from quantities such as the inter-phase transfer areas, mass transfer coefficients, uptake and release rates or times of chemicals into phases such as fish or sediment, and Z values. Methods of calculating D value are described by Mackay and Paterson [3].

Non-diffusive transfer processes may also be included in which the substance is transferred from one compartment to another in association with a volume of material which changes phase. Examples are wet and dry deposition from the atmosphere, sediment deposition and resuspension, and food ingestion by biota. If the volume flow rate is G_i m^3/h and the concentration is C_i then the flux is G_iC_i or $G_iZ_if_i$ or D_if_i where D_i is a transport parameter equal to G_iZ_i. This transfer process takes place in one direction only.

The steady state mass balance for each compartment may be written as

$$E_i + G_iC_{Bi} = f_iV_iZ_i(k_i + k_{Ai}) + \sum D_{ij}f_i - \sum D_{ji}f_j.$$

The simultaneous linear equations thus obtained may be solved by matrix inversion or in some cases by algebraic solution [3].

A Level III calculation which includes expressions for interphase transport resistances, advective flows, and, in the case of mirex, assumes emissions into the water phase of 1 mol/h is illustrated in Figure 4. Each box represents a phase, or compartment, with its corresponding concentrations, amounts, percentages, and fugacities. Interphase transfer, advective flow, and degradative processes are depicted by various arrows, with wider arrows indicating the major processes.

The resulting distribution shows high accumulation of the contaminant ($\sim 99\%$) in the sediment. This is mainly due to a high K_{OW}, causing mirex to sorb to suspended sediment, with subsequent deposition and burial.

This calculation demonstrates that mirex is a highly persistent substance now with an estimated reaction persistence τ_R of 23 years, causing a long term sediment contamination problem. Air advection, is the major removal process from the Unit World producing an overall flow and reaction time τ_O of 5 years, but may only transfer the problem to another geographic area. The Level III model gives a much more accurate picture of the chemical's behavior, especially when the fate is sensitive to the phase into which emissions are introduced.

Level IV Fugacity Model

The Level IV model describes the non-equilibrium, unsteady state behavior of a substance in the environment. It takes the form of a set of differential equations for the compartments with respect to time. They may be written: –

$$V_idC_i/dt = V_iZ_idf_i/dt = E_i - f_iV_iZ_ik_i - \Sigma_jD_{ij}f_i + \Sigma_jD_{ji}f_j.$$

These equations permit time-varying emissions to be considered and the times to build to steady state, or to decay after emission reduction, to be determined. This capability is useful in assessing the possibility of long-term contamination problems for new and existing substances such as mirex, PCBs, dioxins, and mercury. Figure 5 is a schematic illustration of mirex build-up during steady emissions, then decay after emissions cease. After prolonged, steady emissions the concentrations approach the values calculated in Level III.

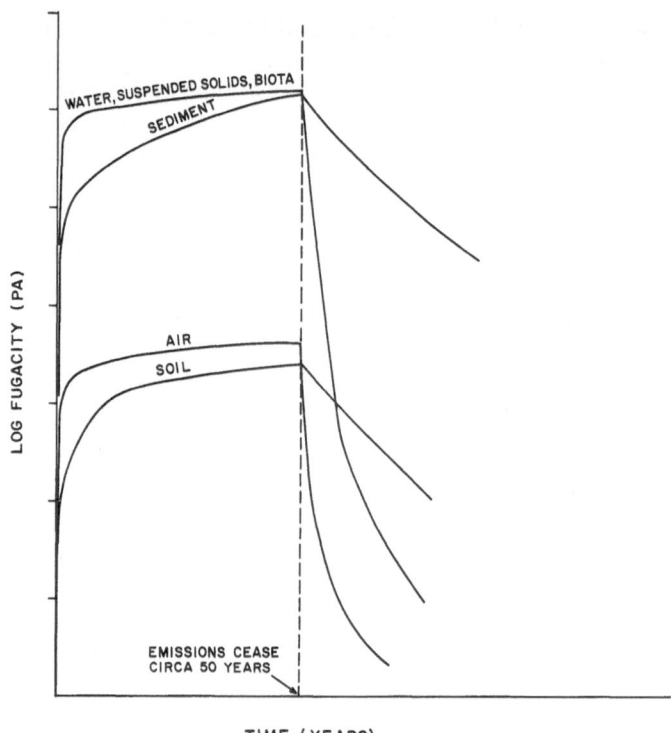

Fig. 5. Schematic plot of fugacity vs time for a level IV calculation for Mirex showing different time responses of each compartment

Level V

Concentrations in a real environment vary spatially and temporally. It is useful to determine how they vary and whether certain fractions of a population are experiencing excessive concentrations, and what this fraction is. Level V considers these heterogeneous distributions as they apply spatially.

Georgopoulos and Seinfeld [17] and others have assessed the temporal variation of concentrations of contaminants in air; and Dean [18] has reported similar studies in water. Recently, Mackay and Paterson [19] have attempted to describe the spatial variation of chemicals in environmental compartments, giving specific examples of mirex and PCB's in sediments. These environmental distribution patterns are usually described by normal, lognormal or Weibull functions.

Each distribution function has two forms, (i) a (usually) bell-shaped distribution function with ordinate y, and (ii) a cumulative distribution function with ordinate F (see Figure 6).

If V_T m^3 is the total volume of an actual or evaluative environmental compartment containing M mols of chemical, then the mean concentration is C_M or M/V_T mol/m^3. However, in the case where the concentration varies within the compartment, y becomes effectively the fraction of the total volume experiencing

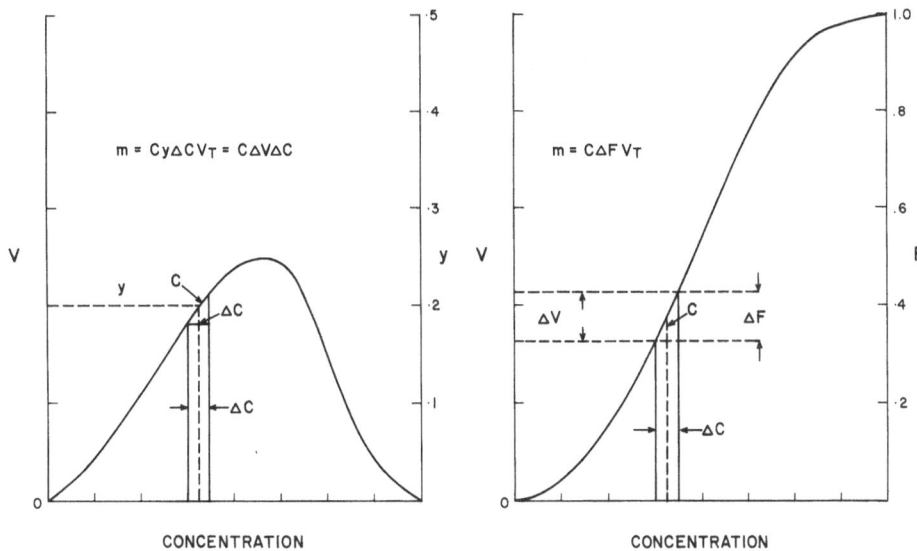

Fig. 6. Schematic illustration of a) a distribution function, and b) a cumulative distribution function showing calculation of amounts in a given concentration range

a concentration in the range C to C + ΔC (mol/m³) and has units of reciprocal concentration. As shown in Figure 6, F is the volume fraction experiencing less than a given concentration and is dimensionless with a value between zero and unity. The distributions are constrained by the limits

$$C \to \infty, \; F \to 1.0, \; y \to O$$

and when

$$C = O, \; F = O.$$

The volume V which has a concentration between C and (C+ΔC) mol/m³ (i.e., ΔC) can be ascertained from the y curve as $V_T y \Delta C$ m³ or from the F curve as $V_T \Delta F$ m³. The amounts can then be determined as $CV_T y \Delta C$ or $CV_T \Delta F$ mols and the total amount of chemical M in the compartment is then $\Sigma CV_T y \Delta C$ or $\Sigma CV_T \Delta F$ and can be expressed by the integral

$$V_T \int_0^\infty (Cy)dC = V_T C_M.$$

From this relationship, one parameter is generally determined which relates y to C. A second parameter determines the degree of spread on either side of the mean concentration. A third parameter may be introduced providing a constraint on the minimum value of C. This is not considered here.

Two methods may be used when applying these distributions to real or evaluative environmental data.

The first method is to plot monitoring data for a known volume V_T, calculate the total amount M, mean concentration C_M and spread factor and obtain an equation with these fitted parameters that satisfies the data. We term this a "forward" process. The second method is to take "evaluative" values of V_T, M, and C_M, and develop a suitable, predictive distribution. For this "reverse" process, some additional information on the expected "spread" of the data is required. If this parameter is not available, it may be approximated from environmental measurements of similar compounds in similar environments.

Three distribution and cumulative functions along with the calculation of their means, medians, and modes are given in Table 2. The reader is referred to the work of Hahn and Shapiro [20], and Aitchison and Brown [21], and Weibull [22] for a full description of these functions. The Weibull function has the advantage that it can be integrated directly to produce the cumulative function. This fact along with the properties that it does not produce negative values and can yield a variety of curves including one similar to the lognormal curve make it a mathematically and environmentally useful function. Much work remains to be done in applying such distribution functions to environmental concentration data.

Fugacity Models of Real Environments

A Quantitative Water-Air-Sediment Interaction (QWASI) Model

In a real contamination situation in a river or lake, the QWASI fugacity model [23] may be applied to estimate the extent of contamination and possible recovery times of an air-water-sediment system.

The model describes a body of water, its partitioning and degradation characteristics and flows of air, water, and suspended sediment into and out of the system. Equilibrium between phases is again described by equal fugacities and partitioning by fugacity capacities, with uniform fugacity prevailing throughout each phase. By defining degradation and transport parameters in identical units, the model facilitates the comparison and mathematical handling of these rates. The calculation may be applied to steady or unsteady state problems. The essential features of the model are described below.

In formulating the model, water and sediment volumes are defined as well as air-water and water-sediment interfacial areas. Fugacity capacities (Z values) are calculated as before. Flow rates of various processes such as advection, sediment and air particle deposition and rainfall are also defined.

Degradation or transformation process rates in water and sediment which were expressed as $V_i C_i k_i$ or $V_i Z_i f_i k_i$ in the evaluative Levels II–IV calculations are expressed here in terms of a D value, or transformation parameter, the rate being given as $D_i f_i$ mol/h where D_i is $V_i Z_i k_i$ mol/Pa.h. An expression is included for diffusive air-water exchange (volatilization and absorption) using a similar approach to that used in Level III evaluative calculations. Diffusion between sediment pore-water and water is also characterised by a D value which is estimated from a mass transfer coefficient or diffusivity for sediment-water exchange. The

Table 2. Distribution functions and their properties (Reprinted with permission from Environ. Sci. & Technol. *18* (7), 1984, American Chemical Society)

	Distribution function	Cumulative function	Mean	Properties Median	Mode
Normal	$\dfrac{1}{\sigma\sqrt{2\Pi}}\exp\left[\dfrac{-(C-C_M)^2}{2\sigma^2}\right]$	$\dfrac{1}{\sqrt{2\Pi}}\displaystyle\int_{-\infty}^{z}\exp\left[\dfrac{-C^2}{2}\right]dC$ where $Z=\dfrac{C-C_M}{\sigma}$	C_M	C_M	C_M
Lognormal	$\dfrac{1}{C\ln S\sqrt{2\Pi}}\exp\left[\dfrac{-(\ln C-\ln C_G)^2}{2(\ln S)^2}\right]$	$\dfrac{1}{\sqrt{2\Pi}}\displaystyle\int_{-\infty}^{z}\exp\left[\dfrac{-C^2}{2}\right]dC$ where $Z=\dfrac{\ln C-\ln C_G}{\ln S}$	$C_G\exp\left[\dfrac{(\ln S)^2}{2}\right]$	C_G	$\dfrac{C_G}{\exp[(\ln S)^2]}$
Weibull	$\dfrac{\lambda}{C_W}\left[\dfrac{C}{C_W}\right]^{\lambda-1}\exp\left[-\left[\dfrac{C}{C_W}\right]^{\lambda}\right]$	$1-\exp\left[-\left[\dfrac{C}{C_W}\right]^{\lambda}\right]$	$C_W\Gamma(1+1/\lambda)$	$C_W(0.693)^{\lambda}$	$C_W\left[\dfrac{(\lambda-1)}{\lambda}\right]^{1/\lambda}$ for $\lambda>1$

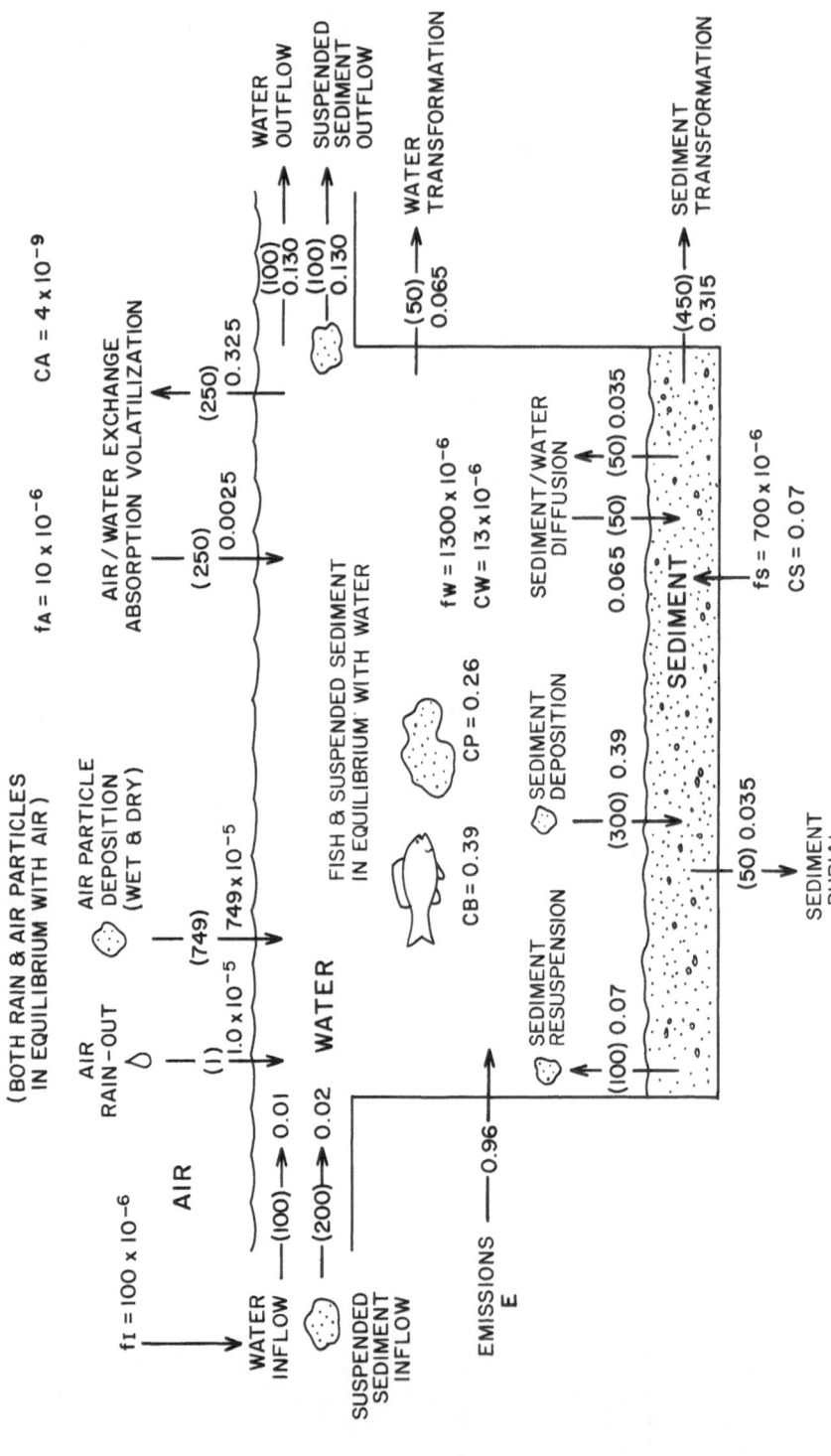

Fig. 7. Steady state solution with D values (in parentheses) fluxes (mol/h) and concentrations (mol/m³)

Table 3. Quasi model processes and rate expressions (Reprinted with permission from Chemosphere *12* (7/8), 1983, Pergamon Press Ltd)

Process		Rates	(mol/h)	D and D group
Sediment burial	G_BC_S	$G_BZ_Sf_S$	D_Bf_S	$D_B=G_BZ_S$
Sediment transformation	$V_SC_Sk_S$	$V_SZ_Sk_Sf_S$	D_Sf_S	$D_S=V_SZ_Sk_S$ ⎤ D_1
Sediment resuspension	G_RC_S	$G_RZ_Sf_S$	D_Rf_S	$D_R=G_RZ_S$
Sediment-to-water diffusion	$K_TA_SC_N$	$K_TA_SZ_Wf_S$	D_Tf_S	$D_T=K_TA_SZ_W$ ⎤ D_2
Water-to-sediment diffusion	$K_TA_SC_W$	$K_TA_SZ_Wf_W$	D_Tf_W	$D_T=K_TA_SZ_W$
Sediment deposition	G_DC_P	$G_DZ_Pf_W$	D_Df_W	$D_D=G_DZ_P$ ⎤ D_3
Water transformation	$V_WC_Wk_W$	$V_WZ_Wk_Wf_W$	D_Wf_W	$D_W=V_WZ_Wk_W$
Water-to-air volatilization	$K_VA_WC_W$	$K_VA_WZ_Wf_W$	D_Vf_W	$D_V=K_VA_WZ_W$ ⎤ D_4
Water outflow	G_JC_W	$G_JZ_Wf_W$	D_Jf_W	$D_J=G_JZ_W$
Suspended sediment outflow	G_YC_P	$G_YZ_Pf_W$	D_Yf_W	$D_Y=G_YZ_P$ ⎤ D_5
Air-to-water absorption	$K_VA_WP_A/H$	$K_VA_WZ_Wf_A$	D_Vf_A	$D_V=K_VA_WZ_W$
Air particle deposition	G_QC_Q	$G_QZ_Qf_A$	D_Qf_A	$D_Q=G_QZ_Q$ ⎤ D_6
Air rain-out	G_MC_M	$G_MZ_Wf_A$	D_Mf_A	$D_M=G_MZ_W$
Water inflow	G_IC_I	$G_IZ_Wf_I$	D_If_I	$D_I=G_IZ_W$
Suspended sediment inflow	G_XC_X	$G_XZ_Pf_I$	D_Xf_I	$D_X=G_XZ_P$ ⎤ D_7

It may be preferable to express certain material transfer rates on an area specific basis $(J\,m^3/m^2 \cdot h)$, namely

$$G_B=J_BA_S, \quad G_R=J_RA_S, \quad G_D=J_DA_S, \quad G_Q=J_QA_W, \quad G_M=J_MA_W$$

fugacities of sediment and its pore-water are assumed to be equal and the diffusion from water to sediment is expressed as D_Tf_W with D_Tf_S defining the reverse process, the net water to sediment flux being $D_T(f_W-f_S)$. Sediment-water transfer processes of sediment resuspension and deposition and sediment burial are expressed in a similar nondiffusive manner with D values defined as GZ products.

Material transport processes between air and water include deposition of the substance in solution in the form of rain or in sorbed form on wet or dry particulate matter. In both cases the rate is quantified as a Df or GZf product thus D is GZ.

Advection into and out of the system by air, water, and suspended sediment are similarly quantified by D values defined as GZ products.

The air fugacity is assumed to be constant (and is often zero) and emissions into the water are defined.

In summary, all transformation and transport process rates are defined by D values, the rate being Df where f is the fugacity of the relevant phase of air, water or sediment. No concentrations are used in the calculations, only fugacities. Figure 7 illustrates these processes for a hypothetical compound and Table 3 gives the definition of the various D values. The advantage of this approach is that certain D values can be added (for example, those for sediment burial and transformation) to give a single D group, thus reducing the algebraic complexity. Mass balance equations can be written in algebraic form describing steady state conditions or in differential form describing unsteady state conditions and solutions obtained for fugacities and hence concentrations. Concentrations in fish and suspended sediment can be estimated by assuming that their fugacities are equal to that of the water.

Details of the equations and illustrative examples are given by Mackay et al. [23, 24] for lakes and sections of rivers.

Human Exposure

The aim of chemical fate calculations is often to assess human exposure. Following the approach of Rosenblatt [25], the model output can be used to estimate the amount of chemical which reaches a "typical" human by various routes based on exposure to the media as shown in Table 4. The extent to which the chemical is actually absorbed (dosage) and the subsequent toxicological effects are not considered here. If an acceptable daily exposure to a toxicant is suggested then corresponding allowable concentrations in crops, animals, fish, drinking water, and air can potentially be determined.

The most difficult medium to quantify is food other than fish, e.g. cereal crops, meat, and vegetables. The extent of contamination of these foodstuffs is very difficult to predict and undoubtedly it varies greatly, especially if the contaminant is a pesticide applied agriculturally. Several approaches are possible such as the assignment of mean lipid and water contents and a fugacity equal to that of the soil or intermediate between soil and air. For purely illustrative purposes and primarily to indicate a future direction for incorporation of models into ex-

Table 4. Illustrative calculation of human chronic exposure. Values for Mirex using exposure estimates of Rosenblatt et al. [25] and assumed fugacity and Z values for foods other than fish

Route	Fugacity (Pa)	Concentration		Exposure	Amount (ng/day)
		(mol/m^3)	(ng/m^3)		
Air	3.72×10^{-9}	1.5×10^{-12}	0.82	18.5 m^3/day	15
Water	1.14×10^{-6}	1.1×10^{-9}	600	2 L/day	1.2
Foods					
Fish	1.14×10^{-6}	4.11×10^{-4}	2.2×10^8	0.032 kg/day	7000
1 Crop[a]	2.67×10^{-9}	1.09×10^{-7}	6.0×10^4	0.063 kg/day	3.8
Meat	2.67×10^{-9}	1.09×10^{-7}	6.0×10^4	0.126 kg/day	7.6

[a] Grain, fruit, root, etc.
f for food $= 2.67 \times 10^{-9}$ Pa
Z for food $= 41$ mol/m$^3 \cdot$ Pa

posure assessment, we define this food as having a Z value corresponding to 10% of Z for fish, plus 80% of Z for water, plus 5% of Z for soil and having a fugacity equal to that of soil. It is emphasized that this is entirely speculative.

The relative mirex exposure amounts calculated for each medium are shown in Table 4. The mirex concentrations in Figure 4 resulting from the Level III calculation are unrealistically high because of the arbitrary selection of unit world volumes and 1.0 mol/h emission rate. More realistic, but only illustrative, concentrations are obtained if an emission rate of 10^{-4} mol/h is assumed, i.e. the concentrations in Figure 4 are multiplied by 10^{-4}. These concentrations and the corresponding exposure amounts are given in Table 4. This information provides an additional dimensions to the behaviour profile by identifying the likely dominant exposure route of mirex which in this case is fish which contains 0.22 g/m^3 or ppm. This is not necessarily the dominant dosage route as it may not be absorbed from the gastro intestinal tract. This requires additional calculation.

Conclusions

In the past decade the science of environmental contaminant chemistry has matured from a mere description of phenomena (e.g., DDT bioconcentrates) to a quantitative treatment of partitioning, reactions, and interphase transport. Many of these phenomena can now be measured in the laboratory under controlled, isolated conditions, then the data extrapolated to environmental conditions in which other complementary and competing processes occur. It is clear that mathematical models, such as the fugacity models, have an important role to play in the process of building a quantitative understanding of environmental chemistry. Such models are a vehicle for synthesis of partitioning, reaction, and transport data into a single comprehensive chemical behavior profile. Without the model, the data tend to be merely a collection of disparate quantities which have different units and can not be mentally assimilated into an overall behavior profile. It is likely that in the next decade several models and modelling systems will evolve, be tested and will be improved. Ultimately, the most reliable and useful will emerge as generally accepted. The fugacity models may be one of these.

The use of fugacity should not be regarded as a replacement for concentration, merely as a supplement which in certain cases facilitates interpretation. Fugacity is enlightening as an "activity" quantity just as pH as an expression of hydrogen ion activity is often more enlightening than total concentration (associated and dissociated) of acid. Chemicals tend to diffuse from regions of high to low fugacity; thus viewing the environment through the lens of fugacity may reveal "hot spots" and conversely regions of constant fugacity such as water and fish which are close to equilibrium, but differ greatly in concentration.

It is hoped that the fugacity calculations described here will be applied to the elucidation of the fate of existing and new chemicals in laboratory systems, to controlled sections of the environment such as ponds, lakes or river sections and to the environment in general. The validity and usefulness of the approach will then be better defined, as will its contribution to the chemistry of environmental contaminants.

References

1. Mackay D.: Finding fugacity feasible. Environ. Sci. & Technol. *13*, 1218 (1979)
2. Mackay, D., Paterson, S.: Calculating fugacity. Environ. Sci. & Technol. *15*(9), 1006–1014 (1981)
3. Mackay, D., Paterson, S.: Fugacity revisited. Environ. Sci. & Technol. *16*, 654–660 (1982)
4. Lewis, G.N.: The law of physico-chemical change. Daedalus, Proc. Am. Acad. *37*, 49 (1901)
5. Prausnitz, J.M.: Molecular thermodynamics of fluid phase equilibrium. Prentice Hall, Englewood Cliffs, N.J. (1973)
6. Van Ness, H.C., Abbott, M.M.: Classical thermodynamics of non electrolyte solutions. McGraw Hill (1982)
7. Mackay, D., Bobra, A., Chan, D., Shiu, W.Y.: Vapor pressure correlations for low volatility environmental chemicals. Environ. Sci. & Technol. *16*, 645–649 (1982)
8. Mackay, D., Bobra, A., Shiu, W.Y.: Relationships between aqueous solubility and octanol-water partition coefficients. Chemosphere *9*, 701–711 (1980)
9. Mackay, D.: Correlation of bioconcentration factors. Environ. Sci. & Technol. *16*, 274–278 (1982)
10. Karickhoff, S.W.: Semi-empirical estimation of sorption of hydrophobic pollutants on natural sediments and soils. Chemosphere *10*, 833–849 (1981)
11. Baughman, G.L., Lassiter, R.R.: *In*: Estimating the hazard of chemical substances to aquatic life. ASTM Tech. Pub. 657, Cairns, J. Jr., Dickson, K.G., Maki, A.W., Eds. 35 (1978)
12. Burns, L.A., Cline, D.M., Lassiter, R.R.: Exposure analysis modeling system (EXAMS): User manual and system documentation. U.S. EPA Environmental Research Laboratory, Athens, GA (1981)
13. Smith, J.H., Mabey, W.R., Bohonos, N., Holt, B.R., Lee, S.S., Chou, T.-W., Bomberger, D.C., Mill, T.: Environmental pathways of selected chemicals in freshwater systems, Vol. II, EPA-600/7-78-074 (1978)
14. Neely, W.B., Mackay, D.: *In*: Modelling the fate of chemicals in the aquatic environment, Dickson, K.L., Maki, A.W., Cairns, J. (eds). Ann Arbor Science, Ann Arbor, MI. 127 (1982)
15. Schmidt-Bleek, F., Haberland, W., Klein, A.W., Caroli, S.: Steps towards environmental haward assessment of new chemicals (including a hazard ranking scheme, based on directive 79/831/EEC). Chemosphere *11*, 383 (1982)
16. Hushon, J.M., Klein, A.W., Strachan, W.J.M., Schmidt-Bleek, F.: Use of OECD premarket data in environmental exposure analysis for new chemicals. Chemosphere *12*, 887 (1983)
17. Georgopoulos, P.G., Seinfeld, J.H.: Statistical distributions of air pollutant concentration. Environ. Sci. & Technol. *7*, 401A (1982)
18. Dean, R.B.: *In*: Chemistry in water reuse, W.J. Cooper (ed). *1*, Ann Arbor Science, Ann Arbor, MI (1981)
19. Mackay, D., Paterson, S.: Spatial concentration distributions. Environ. Sci. & Technol. *18*, 207A (1984)
20. Hahn, G.J., Shapiro, S.S.: Statistical models in engineering. J. Wiley & Sons, Inc. (1967)
21. Aitchison, J., Brown, J.A.C.: The lognormal distribution. Cambridge University Press (1966)
22. Weibull, W.: A statistical distribution function of wide applicability, presented to the American society of mechanical engineers. Atlantic City, N.J. (1951)
23. Mackay, D., Joy, M, Paterson, S.: A quantitative water, air, sediment interaction (QWASI) fugacity model for describing the fate of chemicals in lakes. Chemosphere *12*, 981–997 (1983)
24. Mackay, D., Paterson, S., Joy, M.: A quantitative water, air, sediment interaction (QWASI) fugacity model for describing the fate of chemicals in rivers. Chemosphere *12*, 1193–1208 (1983)
25. Rosenblatt, D.H., Dacre, J.C., Cogley, D.R.: *In*: Environmental risk analysis for chemicals, Conway, R.J., ed. Van Nostrand Reinhold, New York, 474 (1980)

Subject Index

The Handbook
of Environmental Chemistry

Editor: **O. Hutzinger**

Volume 2

Reactions and
Processes

Part A

With contributions by numerous experts
1980. 66 figures, 27 tables. XVIII, 307 pages
ISBN 3-540-09689-2

Contents: Transport and Transformation of Chemicals: A
Perspective. – Transport Processes in Air. – Solubility, Partition Coefficients, Volatility, and Evaporation Rates. –
Adsorption Processes in Soil. – Sedimentation Processes in
the Sea. – Chemical and Photo Oxidation. – Atmospheric
Photochemistry. – Photochemistry at Surfaces and Inter-
phases. – Microbial Metabolism. – Plant Uptake, Transport
and Metabolism. – Metabolism and Distribution by Aquatic
Animals. – Laboratory Microecosystems. – Reaction Types
in the Environment. – Subject Index.

Part B

With contributions by numerous experts
1982. 63 figures. XV, 205 pages
ISBN 3-540-11107-7

Contents: Basic Principles of Environmental Photochemistry. – Experimental Approaches to Environmental
Photochemistry. – Aquatic Photochemistry. – Microbial
Transformation Kinetics of Organic Compounds. – Hydrophobic Interactions in the Aquatic Environment. – Interactions of Humic Substances with Environmental Chemicals.
– Complexing Effects on Behavior of Some Metals. – The
Disposition and Metabolism of Environmental Chemicals
by Mammalia. – Pharmacokinetic Models. – Subject Index.

Springer-Verlag
Berlin
Heidelberg
New York
Tokyo

The Handbook
of Environmental Chemistry

Editor: **O. Hutzinger**

This handbook is the first work that covers the chemical and physical behavior of compounds in the environment.
Under the editorship of Prof. O. Hutzinger, formerly director of the Laboratory of Environmental and Toxicological Chemistry at the University of Amsterdam, now Ecological Chemistry and Geochemistry, University of Bayreuth, 65 international specialists have contributed to Parts A and B of the first three volumes:

– Volume 1: **The Natural Environment and the Biogeochemical Cycles**
– Volume 2: **Reactions and Processes**
– Volume 3: **Anthropogenic Compounds**

For a rapid publication of the material each volume was divided into several parts. Part A of the first three volumes appeared in 1980 and Part B in 1982. Each volume of Part B contains a cumulative subject index. More than 5064 literature references are cited. Future volumes are planned and will cover analytical chemistry, environmental engineering and toxicology.
The Handbook of Environmental Chemistry is a critical and complete outline of our present knowledge and will prove invaluable to environmental scientists, biologists, chemists (biochemists, agricultural and analytical chemists), medical scientists, occupational and environmental hygienists, research geologists and meterologists as well as to industry and administrative bodies.

Springer-Verlag
Berlin
Heidelberg
New York
Tokyo